Moths and Caterpillars

Edited by Vonnie D.C. Shields

Published in London, United Kingdom

IntechOpen

Supporting open minds since 2005

Moths and Caterpillars
http://dx.doi.org/10.5772/intechopen.87469
Edited by Vonnie D.C. Shields

Contributors
Barkat Hussain, Shyam Kumar Vootla, Sayed Iqbal Ahamad, Neeta Kari, Michael Hilary Otim, Girma Hailu, Pamela Paparu, Peter Chinwada, Winnifred Aool Opio, Anani Bruce Yaovi, Allan Obonyom Tekkara, Juliet Akello, Barnabas Mudde, Fiaboe Komi K Mokpokpo, Chandra Pal Singh, Shipra Saxena, Sneha Yogindran, Yogita Sharma, Manmohan Arya, Vonnie D.C. Shields

First published in London, United Kingdom, 2021 by IntechOpen
IntechOpen is the global imprint of INTECHOPEN LIMITED, registered in England and Wales, registration number: 11086078, 5 Princes Gate Court, London, SW7 2QJ, United Kingdom
Printed in Croatia

British Library Cataloguing-in-Publication Data
A catalogue record for this book is available from the British Library

Additional hard and PDF copies can be obtained from orders@intechopen.com

Moths and Caterpillars
Edited by Vonnie D.C. Shields
p. cm.
Print ISBN 978-1-83968-517-0
Online ISBN 978-1-83968-518-7
eBook (PDF) ISBN 978-1-83968-519-4

For EU product safety concerns:
IN TECH d.o.o., Prolaz Marije Krucifikse Kozulić 3, 51000 Rijeka, Croatia, info@intechopen.com or visit our website at intechopen.com.

Meet the editor

Vonnie D.C. Shields, Ph.D., is Associate Dean, Fisher College of Science and Mathematics and a full professor in the Biological Sciences Department, Towson University, Towson, Maryland, USA. Dr. Shields' research explores gustatory, olfactory, and visual cues in insects. Her laboratory employs morphological, behavioral, and electrophysiological techniques to better understand sensory mechanisms by which larval and adult insects find host plants and detect plant-associated volatiles. Dr. Shields received a BS and Ph.D. from the University of Regina, Regina, Saskatchewan, Canada. A portion of her Ph.D. studies was carried out at the University of Alberta, Edmonton, Alberta, Canada. After graduating, she accepted a research associate position to conduct postdoctoral studies at the Arizona Research Laboratories Division of Neurobiology, University of Arizona, Tucson, Arizona, USA, before she joined the faculty at Towson University where she rose through the ranks from assistant to full professor.

Contents

Preface

This book is written by experts in their respective fields and is an invaluable resource for entomologists as well as biologists, ecologists, zoologists, teachers, and students. The topics presented in this book include information regarding (1) the biology, spread, management, and biocontrol of fall armyworm, *Spodoptera frugiperda*, in Africa, (2) RNAi-mediated control of lepidopterous crop pests, (3) use of pheromonal-baited traps to disrupt mating for use against the codling moth, *Cydia pomonella*, (4) the importance of fibroin and sericin, two types of proteins, in the structure and composition of cocoons in polyphagous Tasar silkworm moth *Antheraea mylitta* in India, and (5) how the sense of taste plays a critical role in the feeding behavior of insects, with specific reference to gypsy moth caterpillars, *Lymantria dispar*.

In Chapter 1, "Managing a Transboundary Pest: The Fall Armyworm on Maize in Africa", Otim et al. review efforts to manage the fall armyworm, *Spodoptera frugiperda*, a main pest of maize. This chapter covers the biology, origin, and distribution of this insect species. In addition, the chapter discusses favored host plants, damage, crop losses surveillance and monitoring using pheromonal lures and traps, and field assessments. The authors discuss agroecological management strategies, including weed manipulation and intercropping, transgenic approaches, and chemical control. The authors emphasize the importance of collation and dissemination of research outputs, capacity building efforts, and effective policy implementation as additional strategies for pest and disease management.

In Chapter 2, "RNAi-Mediated Control of Lepidopteran Pests of Important Crop Plants", Saxena et al. examine alternative, more environmentally favorable approaches of insect control in lieu of the use of insecticides and *Bacillus thuringiensis* toxins to improve plant protection. The post-transcriptional gene-silencing mechanism ribonucleic acid interference (RNAi) has surfaced as a new, sustainable, and environmentally favorable approach that supports insect management and crop protection. This latter technology relies on the idea that small pieces of RNA can terminate protein translation by binding to the messenger RNAs that code for those proteins. Small interfering RNA (siRNA) and microRNA (miRNA), in addition to the RNA-induced silencing complex (RISC), bind to the complementary mRNA and prevent ribosomes from continuing to produce the associated protein, thus inducing gene silencing at the post-transcriptional level. RNAi shows much promise as a tool in molecular biology. The authors evaluate how the RNAi mechanism has been applied in controlling lepidopteran crop pests.

In Chapter 3, "Role of Pheromone Application Technology for the Management of Codling Moth in High Altitude and Cold Arid Region of Ladakh", Hussain et al. analyze the use of pheromone dispensers and pheromone-baited traps to disrupt mating for use against the codling moth, *Cydia pomonella*. Annually, this pest is known to destroy various fruit crops, including apples, in various regions of India. While insecticides are cheaper and can target a broad array of insects, they are typically hazardous to consumers, not environmentally friendly, and may lead to resistance and death of natural enemies. The authors consider the constraints involved with current management practices and their limitations with respect to maintaining orchards (i.e., plant pruning, orchard maintenance, climate conditions, etc.) and discuss the benefits of pheromone-driven technology as an important means to manage the codling moth.

In Chapter 4, "Bioactive Secondary Metabolites of Wild *Antheraea mylitta* Silkworm Cocoons", Ahamad et al. examine the chemical compounds, proteins, and secondary metabolites that comprise the unique cocoons of the wild tropical silk-spinning polyphagous Indian Tasar silkworm, *Antheraea mylitta*. More specifically, the authors focus on the importance of two types of proteins, fibroin and sericin, comprising the structure and composition of cocoons that house and protect the pupae, which has allowed this insect species to endure harsh environmental conditions and survive for millions of years. Of interest is the fact that these silk proteins are biologically versatile molecules and have served commercially in the manufacturing of suture material for wound applications and non-load and load-bearing tissue engineering purposes, drug delivery, and other medical applications. Additionally, the authors evaluate the importance of secondary metabolites (e.g., alkaloids, saponins, steroids, phenols, flavonoids, terpenoids, tannins, fatty acids, carboxylic acids, aldehydes, sterols, etc.) that are sequestered from plants by the larvae and their role in cocoon formation.

In Chapter 5, "Functional Morphology of Gustatory Organs in Caterpillars", Shields provides a detailed account of how the sense of taste plays a critical role in the feeding behavior of insects, specifically caterpillars. The author highlights how taste stimuli are recognized, coded, and processed by receptor cells housed in two specific sensory organs (sensilla), the medial and lateral styloconic sensilla. These sensilla are deemed to be the primary organs involved in feeding as they are in continuous contact with plant sap comprised of different phytochemicals, during feeding. Each sensillum houses four gustatory receptor cells and forms a sensory filter for taste signals. An ultrastructural morphological description, using both scanning electron microscopy and transmission electron microscopy, is provided for both styloconic sensilla. In addition, the author examines feeding behavior, phytochemicals, hostplant preferences, and neurophysiological responses of sensory organs involved in peripheral gustatory coding. Specific examples are made to gypsy moth caterpillars, *Lymantria dispar*. The author concludes the chapter by mentioning current molecular methods that are being used to advance the field.

I wish to thank IntechOpen for initiating this book project and inviting me to serve as the main academic editor. I would like to acknowledge Publishing Process Manager Maja Bozicevicfor guiding me through the publication process. I would like to thank all the authors who contributed to this book for their hard work in submitting and editing their contributions. Lastly, I would like to thank an anonymous reviewer for revising the chapter that I contributed, as well as my husband, Dr. Thomas Heinbockel, Professor and Interim Chairperson, Department of Anatomy, Howard University College of Medicine, and our son, Torben Heinbockel, for their patience and understanding when I was working on this book project.

Dr. Vonnie Denise Christine Shields
Associate Dean,
Jess and Mildred Fisher College of Science and Mathematics,
Towson University,
Towson, Maryland, USA

Professor,
Biological Sciences Department,
Towson University,
Towson, Maryland, USA

Chapter 1

Managing a Transboundary Pest: The Fall Armyworm on Maize in Africa

Michael Hilary Otim, Komi Kouma Mokpokpo Fiaboe, Juliet Akello, Barnabas Mudde, Allan Tekkara Obonyom, Anani Yaovi Bruce, Winnifred Aool Opio, Peter Chinwada, Girma Hailu and Pamela Paparu

Abstract

The fall armyworm (*Spodoptera frugiperda* J.E Smith) (Lepidoptera: Noctuidae) invaded Africa in 2016, and has since spread to all countries in sub-Saharan Africa, causing devastating effects on mainly maize and sorghum. The rapid spread of this pest is aided by its high reproductive rate, high migration ability, wide host range and adaptability to different environments, among others. Since its introduction, many governments purchased and distributed pesticides for emergency control, with minimal regard to their efficacy. In this chapter, we review efforts towards managing this pest, highlight key challenges, and provide our thoughts on considerations for sustainable management of the pest.

Keywords: agroecology, parasitoids, pesticides, *Spodoptera frugiperda*, *Zea mays*

1. Introduction

The fall armyworm (FAW) *Spodoptera frugiperda* (J.E Smith) (Lepidoptera: Noctuidae) was first reported in Africa in 2016 [1], where it mainly impacts maize production. Annual yield losses in maize due to FAW infestation are estimated between 8.3 and 20.6 million tons in just 12 African countries, valued at US$2,481–6,187 million [2]. When the FAW was reported in Africa in 2016 [1], there was a general lack of knowledge on its management, causing a panic that resulted in many African governments procuring and distributing non-validated insecticides for its control. At this time, many African scientists relied mainly on information and experiences from the management of FAW in the Americas. If left unattended, the continued destruction by the FAW, leading to reduced yields, would aggravate the already precarious conditions of over 400 million Africans living below the poverty line [3]. The impact of FAW would be much felt in Africa with an ever-increasing population and its demand for maize, a preferred food for the poor [4, 5]. The impact of FAW at the household level may not affect the amount of maize consumed, but rather the amount sold because farmers mostly sell off the excess harvest after catering for household food demands. This may affect the income

IntechOpen

earned and result in cash shortages, leading to failure to afford basic necessities [4]. Since its appearance on the continent, substantial research has been conducted on FAW in different African countries, resulting in several publications. The objective of our review is therefore to provide an overview of the management of FAW on maize in Africa, challenges faced and thoughts towards sustainable management of the pest on the continent.

2. Origin and distribution

The FAW is native to the tropical and sub-tropical regions of the Americas [1]. It was first described in 1797 in the state of Georgia, USA. It remained a pest in the Americas until 2016 when it was first reported in West Africa (Nigeria), and the island of São Tomé and Principe [1], and subsequently in almost all sub-Saharan African countries within one year [6–11]. To date, it is known to occur in five continents (Africa, Asia, Australia, North and South America). The FAW possesses a great potential to cover wide geographical locations in a short period [12]. The rapid spread of the pest is attributed to its high reproductive capacity, high migration ability and a wide host range.

3. Biology of the fall armyworm

The FAW undergoes complete metamorphosis (egg, larva, pupa and adult) (**Figure 1**). Under optimal conditions, the development of FAW takes approximately 30 days. The eggs are laid on leaves in batches containing 100–200 eggs [13]. Each female can lay an average of 1500 eggs, with a maximum of over 2,000 eggs in a life time [14] . Eggs hatch in two to three days during summer [14]. There are six larval instars with a development duration of two to three weeks, with the last instar being most destructive (causing 70% of FAW damage) [15]. The FAW generally pupates in the soil at a depth of 2 to 8 cm in a cocoon constructed by tying together particles of soil with silk [14]. The pupal stage lasts about eight to nine days

Figure 1.
Developmental stages of the FAW. (a) Adult male (b) adult female, (c) an egg batch (c) and (d) a mature larva. Photo credit: Dr. Girma Hailu.

during summers [14]. Adults are nocturnal, and most active during warm, humid evenings. The adults live for about, but lifespan ranges between 7 to 21 days [14].

Presently, there are two major strains of FAW known in both the native and invasive range [16, 17]. These are the 'corn' (C strain) and the 'rice' (R strain) strains, which are defined by the host where they develop. The corn strain attacks corn/maize, sorghum and cottons, while the rice strain attacks grasses, like rice, silk grass and some forage grasses. The two strains are morphologically identical but differ in genetics, physiology and behavioral characteristics, such as mating and resistance to insecticides [18]. Both strains are present on the African continent [17]. In addition to these strains, there exist mutant strains with resistance to different insecticide classes in both the Americas [19–21] and Africa [22, 23].

4. Host plants, damage and losses caused in maize

The FAW is a polyphagous pest reported to feed on at least 353 crop and non-crop species from 76 plant families [24], with a high preference for the Gramineae family [25]. Farmed grasses like maize, rice, sugarcane, and sorghum are recorded as major hosts whereas dicotyledonous vegetables and cotton are documented as minor hosts [26]. The number of hosts is likely to increase as this pest expands into newer areas [27–29]. Although it is polyphagous, FAW prefers to feed on maize causing substantial yield loss in SSA [2].

The larvae damage plant leaves, stems, branches, and reproductive organs, such as flowers and fruits. The damage by first instar larvae on grasses such as maize appear as silvery patches called the "windowpanes" because one side of the leaf is eaten, leaving the opposite epidermal layer intact. Damage by the third and fourth instar larvae is more pronounced with holes appearing on the edges going inwards and the characteristic row of perforations due to feeding on the whorl of the growing plants are visible. The larvae also migrate from the leaves to the tassels and the developing ears/grains leading to grain yield losses and exposing the grains to mycotoxin contamination (**Figure 2**). Commonly observed in maize is reduced plant stand in young crops and defoliation of older plants [30].

The losses associated with FAW vary with factors such as host species, varieties, environmental conditions, and socio-economic factors. In maize, FAW infestation

Figure 2.
Symptoms of maize foliage damage. And inset damage to a cob at maturity.

starts from the seedling stage till the reproductive stages (**Figure 2**), causing quantitative and qualitative grain losses. The qualitative losses involve contamination with aflatoxins and fumonisins [31]. In Africa, maize losses were estimated to be 8.3 to 20.6 million tons, causing annual losses of US$2,481–6,187 million [2]. Yield losses is selected African countries are shown in **Table 1**.

Beside the direct effects on yields, FAW infestation may result in significant expenditures on insecticides for both the farmers and governments, and detrimental effects of residual chemicals on human health and the environment [4, 32]. The reduced yields may affect other value chain actors such as livestock farmers who may suffer reduced quantities and quality of maize stover, maize-based food processors who may suffer reduced volumes leading to reduced trade.

Country	Percentage yield losses (%)	References
Benin	42.8 – 59.5	[32, 33]
Ethiopia	11.9	[4]
Ghana	0-40	[34]
Kenya	32-34	[4]
South Africa	26.5 – 56.8	[35]
Tanzania	10.8	[36]
Uganda	0 - 50	Otim et al, unpublished
Zambia	0-50	[34, 37, 38]
Zimbabwe	11.57	[39]

Table 1.
Yield losses caused by FAW in selected countries in sub-Saharan Africa

5. FAW surveillance, monitoring and field assessments

Surveillance at the farm level is an informal and passive way of detecting pest problems as they emerge [40]. Monitoring on the other hand is an exercise that actively tracks the presence and movement of a pest within a given area [40] and is often organized and implemented at various scales, mostly by governments, through trained technical personnel who collect data according to prescribed sampling protocols. Data collected is used to make decision on management of the pest.

In 2019, FAO and CABI advocated for FAW monitoring using commercial pheromone lures and traps to give advance warning to farmers at the beginning of the maize cropping season. To promote harmonization of data collection and reporting procedures as well as foster collaboration among regional member countries, FAO developed a Fall Armyworm Monitoring and Early Warning System (FAMEWS) mobile application tool which requires users to input both field scouting and pheromone trap data [13]. The application tool also helps farmers to correctly identify FAW while providing them with off-line, free control advice delivered via satellites [41].

In addition to FAMEWS, FAO and Pennsylvania State University jointly developed and launched an innovative talking mobile app called *Nuru* (Swahili for "light") in several African countries [13]. *Nuru* is an innovative talking app that uses cutting-edge technologies involving machine learning and artificial intelligence. *Nuru* helps farmers recognize FAW and take immediate measures to destroy it. It runs inside a standard Android phone and can also work offline. *Nuru* is embedded in the *PlantVillage* app, which is a free app built by FAO, CGIAR and other public institutions at Pennsylvania State University and is proposed to be linked into FAO's FAMEWS app. This platform analyses data from all submitting countries across Africa and generates real-time maps giving an overview of the FAW infestations, severities and the measures applied to reduce its impact.

Monitoring of adult FAW using sex pheromone traps (attractive to male moths) is informative of the presence of the pest within a location. Experiments conducted in French Guiana [42] showed that FAW pheromone trap data can be used to estimate, a week in advance, the subsequent abundance of larvae in pastures. However, the above prediction method has been found to be unreliable in crops. For example, McGrath et [40] observed no relationship between the number of FAW male moths caught in traps and the number of female moths laying eggs in the same locality. The foregoing can be attributed to FAW's pre-oviposition period of 3–4 days and migratory tendencies, which therefore means that populations of egg-laying moths in an area may be dominated by migrants which previously mated elsewhere. Thus, catches of male moths in traps should simply be used to estimate the presence of potential egg-laying females in the area.

FAW's lack of diapause combined with a conducive climate in most countries of SSA obviously mean that there are always some populations which do not undertake long-distance migrations as they simply shift to off-season irrigated maize. In the context of managing resistance at a national or regional level, surveillance and monitoring programs can be very useful in mapping the dispersal of insecticide resistant FAW populations. Using an appropriate mark-recapture approach as that used by Osborne et al. (e.g., [43], it should be possible to identify the invasion zones of an insecticide-resistant FAW population arising from a known area. Guidelines on what pesticides to use or not to use in these invasion zones can be issued to curtail the further buildup of selection pressure for resistance development.

Four most used parameters in a FAW field assessments include: [1] pest incidence (% field infestation), [2] plant damage (based on a visual scoring system), [3] pest density per plant or field and [4] yield. For scoring of damage, the Davis and Williams [44] 0–9 visual rating scale is the most commonly used in sub-Saharan Africa [45–47]. However, other researchers have collected FAW damage data using modified versions of the above visual scale. For example, Caniço et al. [40] from Mozambique assessed plant damage in the field using a scale of 0–5, where 0 equals plants with no visual foliar damage, and 5 plants with more than 75% foliar damage or dead from FAW. For evaluating the effectiveness of Bt maize and insecticides for FAW control in Brazil, Burtlet et al. [48] collected damage data based on a visual scale [42] and converted the scales to percentage damages. Maize yield losses due to FAW attack have also been estimated using a variety of protocols, for example the digital imaging method [49]. Rodriguez-del-Bosque et al. [50] compared the weights of damaged kernels per cob with the weight of the same number of undamaged kernels [50]. Based on these few examples, standardization of data collection and analysis protocols is needed to enable comparison of results across regions and countries.

6. Management of the FAW

The methods advocated for controlling the FAW in Africa are agroecological/ cultural, biological control, host plant resistance, transgenic approaches and chemical pesticide use [26]. Below is a review of the use of the above methods on the continent.

6.1 Agroecological management

Agroecological management is the science of applying ecological principles to enhance productivity while reducing the negative impacts on the environment [51, 52]. Successful agroecological management is based on the understanding of key principles such as complex interactions within the ecosystem and contextualized solutions to local problems [53]. Agroecological management practices to reduce FAW populations in Africa include cultural and mechanical control, intercropping, crop rotation with non-host crop, weed management and intercropping maize with the moth repellant Greenleaf desmodium ("Push") with *Brachiaria* cv Mulato planted around the intercrop ("Pull") [53–55].

6.1.1 Cultural and mechanical control of FAW

Cultural control methods for FAW include removal of crop residues and no-tillage. A significant reduction of FAW infestation on maize was observed in maize farms under minimum or no-tillage in Zimbabwe [49]. Minimum tillage also enhances activities of natural enemies. Among the mechanical control methods recommended for and used smallholder farmers in Africa is squashing egg masses and hand-picking small larvae [4, 39, 56]. Farmers in different parts of Africa also resorted to applying sand, ash, or soil in the maize whorl [49, 56, 57]. However ash, soil, and alata samina soap (made from the ash of the barks of plants that are locally harvested, such as plantain, palm tree leaves, and shea tree bark) [58] treatments were found not to be effective in reducing FAW larvae numbers or crop damage at the dosages tested, and thus did not significantly increase maize yields [58].

6.1.2 Weed manipulation

There are mixed observations about the influence of weeds on the population and damage by FAW. For example, Altieri [59] observed significantly less infestation of maize due to FAW in weedy (natural weed complex or selected weed associations) plots compared to weeded plots. Furthermore, in the weedy farm of maize, significantly greater number of FAW predators were encountered. On the contrary, in Zambia, the incidence of FAW was low in frequently weeded plots that were dominated by graminaceous spp. [49]. Bearing in mind the crop-weed competition effect, allowing weeds other than graminaceous in between maize rows and as guard rows enhances the population of natural enemies [60]. Despite the beneficial effect of weeds on the population of arthropod pests, their infestation could cause about 20–50% yield losses in maize [61]. Thus, striking a balance in keeping and removing weeds is important in ensuring high farm productivity.

6.1.3 Intercropping

Intercropping (**Figure 3**) practiced widely by smallholder farmers in sub-Saharan Africa has long been recognized as an efficient farming system providing improved resource utilization and increased productivity [62, 63]. The practice is reported to reduce pest populations and enhance the potential of their natural enemies [64, 65]. In Latin America, maize-bean intercrops reportedly reduced FAW infestation when compared to a maize monocrop [59]. Similarly, studies in Uganda and Cameroon have demonstrated that intercropping maize with beans or groundnuts significantly reduces FAW infestation and damage severity in maize (**Figure 3**) [40, 43]. In Cameroon, a maize intercrop with climbing beans resulted in higher reduction of FAW numbers, compared to bush beans [66]. In the above

Figure 3.
Maize intercropped with groundnuts and common bean in Kamuli, district, Uganda.

studies, the yield of maize in the intercrop increased by almost two-fold, compared to the monocrop. However, the effect of intercropping on grain yield was not comparable to that of synthetic pesticides where a three-fold yield increment was observed. Although further studies are required to determine the mechanism by which intercropping reduces damage caused by the FAW, barrier effect, repellant volatiles emission and enhanced natural enemy abundances are speculated to be key mechanisms [67].

6.1.4 Planting dates of maize in intercrops and monocrops

Early planting of companion crops in a maize cropping system seems to provide masking effect resulting in reduced FAW infestation. For example, a study where simultaneous planting of maize with beans and sequential planting where maize was planted after 20 to 30 days resulted in a significantly less FAW infestation compared to simultaneous cropping [59]. Similar observation was made in maize under push-pull technology where the perennial desmodium sprouted prior to the maize germination. It is however observed that sole maize planted early equally suffers less damage compared to late-planted maize (Otim, pers. Obsv).

6.1.5 Conservation agriculture

The basic principles around conservation agriculture include minimum tillage, crop rotation, and cover crop or mulching [68]. In no-tillage plots where crop residue from the previous harvest was applied as mulch, oviposition and damage by FAW was significantly lower in maize at 2 to 3 leaf stage compared to the plowed plots [69]. The masking effect of the mulch helps maize to escape the early infestation of FAW. Although the level of infestation was less, the effect of the mulch was at par with the plowed farm 20 days after planting, implying the masking effect. Findings from Zambia showed significantly less FAW infestation in maize under zero tillage followed by minimum tillage [49].

6.1.6 Push-pull technology

A novel agricultural technology based on cereal/legume intercropping was developed to tackle multiple problems including insect pests and weeds while augmenting soils with nutrients. The technology was developed by the International Center of Insect Physiology and Ecology (*icipe*) in collaboration with Rothamsted Research in the UK and national partners in east Africa. This technology used stimulo-deterrent diversionary tactic to repel gravid moths of cereal stemborers and FAW from maize due to the intercropped desmodium (push) while attracting them

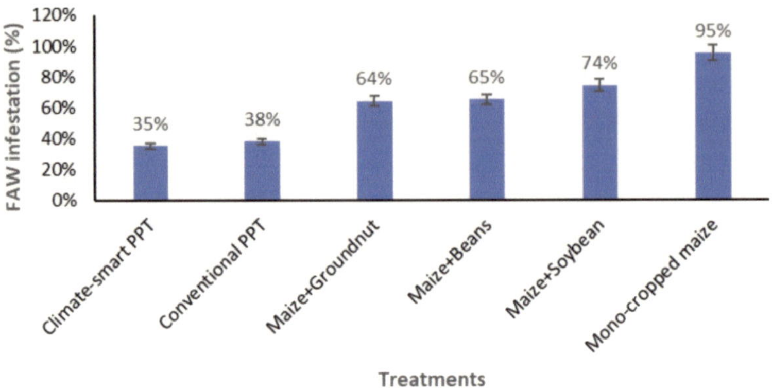

Figure 4.
Percentage fall armyworm infestation of maize in a monocrop, intercrops and under the push-pull technology in Uganda [47].

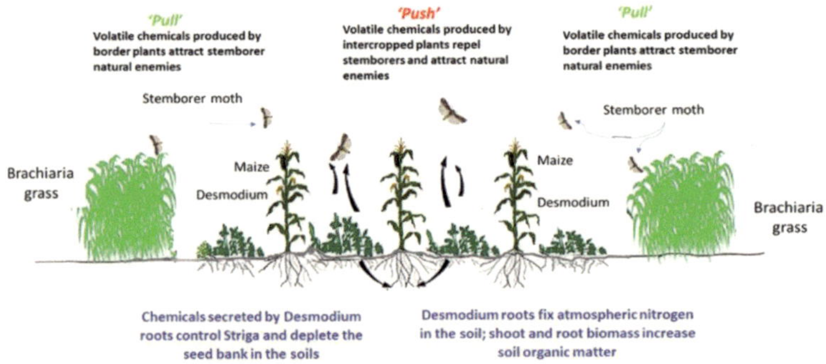

Figure 5.
Climate-smart push-pull technology: Maize intercropped with repellent green leaf desmodium and Brachiaria grass as border crops.

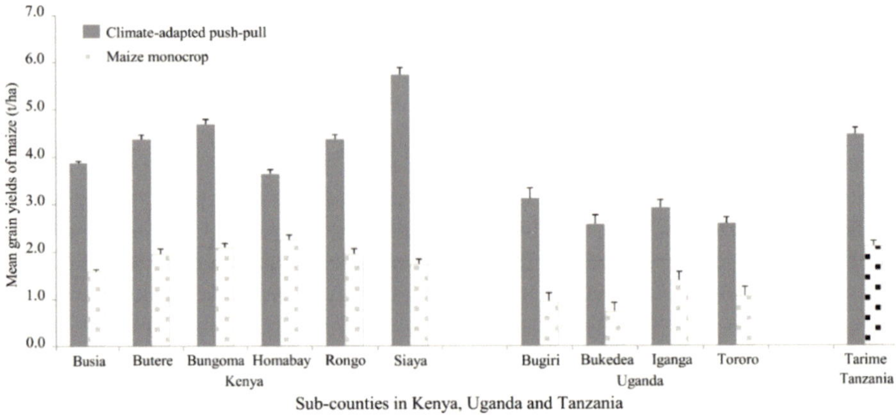

Figure 6.
Mean (± S.E.) grain yields of maize (t/ha) planted in sole stands (maize monocrop) or in climate-adapted push–pull stands [71].

to the trap companion plants such as Brachiaria and Napier grass (pull) planted around the maize plots [70] (**Figure 4**). There are two types of push-pull technologies conventional (maize intercropped with silverleaf desmodium and Napier grass planted around the farm) and climate-smart a climate adapted technology where maize is intercropped with green leaf desmodium and the plot surrounded by Brachiaria (Mulato II) grass (**Figure 5**) [70].

In a study conducted by Midega et al. [55] in Kenya, Uganda, and Tanzania, the team reported 82.7% and 86.7% reduction in larvae per plant and plant damage, respectively in climate-adapted push-pull compared to maize monocrop plots, and a resultant 2.7-fold higher grain yield in the climate-adapted push-pull plots (**Figure 6**). There was a similar finding in Uganda, where maize yield from climate-adapted PPT was significantly greater compared with maize intercropped with edible legumes or mono-cropped maize [54].

6.2 Biological control of FAW in Africa

Centuries of research on FAW in the Americas have recognized the importance of biological control in the management of FAW in its native range. Biological control does not only offer a sustainable solution to FAW management, but also presents an economically and environmentally safer alternative or complement to synthetic pesticides as well as a vital solution for mitigating potential pesticide resistance that is often associated with inappropriate pesticide use [19]. The key categories of natural enemies of FAW are predators, parasitoids, entomopathogens (fungi, bacteria, viruses and nematodes). In Africa, efforts are underway to identify the natural enemies of FAW, assess their impacts and harness their use in the overall management of the pest.

6.2.1 Natural enemies of FAW reported in Africa

Studies on the identification of natural enemies of the FAW in Africa have been reported from different countries including Benin, Cameroon, Cote d'Ivoire, Ethiopia, Ghana, Kenya, Mali, Mozambique, Niger, South Africa, Tanzania, and Uganda. Most of the natural enemies of FAW reported since the advent of the pest on the continent are parasitoids. These are mainly Dipterans and Hymenopterans and were found on either egg, larvae, or pupae of the pest. Highest diversity was found among larval parasitoids, and included *Anatrichus erinaceus* Loew, *Charops ater* Szépligeti, *Chelonus bifoveolatus* Szépligeti, *Coccygidium luteum* Brullé, *Cotesia icipe* Fernandez-Triana & Fiaboe, *Drino quadrizonula* Thomson, *Meteoridea testacea* Granger, *Metopius discolor* Tosquinet, *Palexorista zonata* Curran, *Pristomerus pallidus* Kriechbaumer, and *Procerochasmias nigromaculatus* Cameron [72–78]. Other parasitoid species belonging to the Ichneumonidae (i.e. ichneumonids) and Tachinidae (i.e. tachinids) families were, however, observed in maize and sorghum fields during surveillance studies conducted between 2017 and 2018 [72]. The most promising parasitoids, judged through their widespread distribution on the continent and high parasitism rates, are the egg parasitoids: *Chelonus curvimaculatus* (Cameron), *T. remus*, *Trichogramma* sp. and *Trichogrammatoidea* sp. (both Hymenoptera: Trichogrammatidae) [72, 74, 75, 79]. The occurrence of *T. remus* for instance was reported in Benin, Cameroon, Côte d'Ivoire, Kenya, Niger and South Africa. On the other hand, species in the genus *Chelonus* that were reported to have wide geographical distribution in the Americas [80] and Australia [81], has equally been documented in some parts of Africa (Uganda, Benin, Ghana, etc). In the case of egg parasitoids parasitism of up to 64% were observed in Niger following augmentative release of *T. remus* in sorghum fields [82]. Larval parasitism is still low averaging 9.2 and 9.5% in Uganda and Mozambique, respectively [76, 78].

Few studies have paid attention to predatory species on FAW in Africa. Koffi et al. [77] recorded *Pheidole megacephala* (F.) (Hymenoptera: Formicidae), *Haematochares obscuripennis* Stål, and *Peprius nodulipes* Signoret (both Heteroptera:

Reduviidae) in Ghana. From a field survey conducted in Ghana, *Cheilomenes lunata* Fabricius (Coleoptera: Coccinellidae) and *Ropalidia fasciata* Fabricius (Hymenoptera: Vespidae) were highlighted as predators on FAW larvae, with *R. fasciata* being more recurrent from field observations [83].

Since FAW was reported in Africa, limited research reports exist on entomopathogens of FAW. Akutse et al. [78, 84–86] demonstrated ovicidal and larvicidal potency of isolates from *Metarhizium anisopliae* and *B. bassiana* under laboratory conditions in Kenya. Promising efficacy of *B. thuringiensis* serotype *kurstaki* and *Pieris rapae* granulovirus based formulations in laboratory and field conditions were reported in Ghana [79]. In Tanzania, attempts have equally been made to integrate *M. anisopliae* and *B. bassiana* into diverse cropping systems for FAW management [80].

6.3 Use of botanical against fall armyworm in Africa

In Ethiopia, Sisay et al. [47] associated more than 90% larval mortality to botanical insecticides, namely *Azadirachta indica* Juss, *Schinnus mole* L. and *Phytolacca dodecandra* L'Her under greenhouse conditions.

The ever-increasing diversity of natural enemies being reported across the continent calls for an intensification in the search for local natural enemies coupled with their conservation and mass rearing for conservative and augmentative biological control programs. Priority should be given to this approach over classical biological control and more efforts are required across the continent for prospection for natural enemies, rearing, performance assessment, mass production and release. Efficacy trials are being conducted in various countries for botanicals and biopesticides. Critical aspect hindering their adoption might be the comparatively higher costs compared to chemical pesticides. Thus, strong advocacy activities are required to engage policy makers in establishing conducive environment for wider use of biological control products over chemical pesticides.

6.4 Host plant resistance for managing fall armyworm

Host plant resistance (HPR) is a cornerstone for any pest management strategy. The use of insect-resistant crop varieties as a component of IPM arises from the ecological compatibility and compatibility with other direct control tactics [87]. HPR works in synergy with biological, cultural, chemical and agroecological practices and works with synergy [88]. HPR is very specific to the target pest or group of pests and does not affect the non-target organisms. HPR is also very persistent throughout the cropping season. The quantitative or polygenic nature of native genetic resistance also offers the opportunity to minimize selection pressure on FAW and prevents emergence of new resistant strains. Moreover, HPR does not involve any additional cost to the farming community the farmers do not need any training and the scaling up with be easily adopted in the farming community in Africa.

The FAW, an invasive pest which recently invaded Africa has evolved with wild maize and later with domesticated maize in Latin America. It is expected therefore that these constant interactions resulted in some degree of genetic adaptation. The genetic resistance to FAW in some plants including maize was available since 1990's [89]. The United States Department of Agriculture (USDA) Agricultural Research Service (USDA-ARS, Mississippi) developed and released a series of maize inbred lines with resistance to FAW including Mp496, Mp701–708, Mp713, Mp714, and Mp716 [87, 90–95], derived primarily from germplasm held by the International Maize and Wheat Improvement Center (CIMMYT, Mexico). CIMMYT developed populations from tropical and subtropical maize inbred lines, CML59–74 and CML121–127, from USDA-ARS germplasm, with FAW resistance [96].

Since the invasion of FAW in Africa, intensive and precise screening of maize germplasm against FAW under artificial infestation by CIMMYT has been ongoing in Kiboko, Kenya [97]. The germplasm screened include CIMMYT Multiple Borer Resistance (MBR) and Multiple Insect Resistance Tropical (MIRT), germplasm developed during Insect Resistant Maize for Africa (IRMA) project, USDA Mississippi germplasm and subtropical elite germplasm. CIMMYT scientists have developed within two years several inbred lines and more than 200 single cross hybrids. The first generation of FAW hybrids were announced on the 23rd of December 2020 [98], including three FAW tolerant hybrids (FAWTH2001–2003), which passed the test of screening in the screenhouse, and later under natural infestation in both on station and on farm fields. The next step will be the national performance trials, variety release and registration by private sector (seed company) for the deployment in target geography in eastern and southern Africa. The same exercise is being done by IITA in West Africa to develop FAW maize resistant varieties and in Southern Africa by ICRISAT on sorghum. A lot of NARs in sub-Saharan Africa (especially in Uganda, Malawi) have also initiated germplasm screening to identify tolerant/resistance materials.

6.5 Transgenic approach to control FAW in Africa

The first transgenic Bt plants were introduced on the market in 1996 [99], whilst the first Bt maize for FAW control was introduced in the USA and Pueto Rico in 2013 [100]. On the continent, Bt maize is currently commercially available only in South Africa, where regulatory authorities have overseen multiple approvals, with more than 20 years of deployment of such products. The MON810 event, which has been cultivated in South Africa since 1997 for the stem borer control, also confers partial resistance (50%) to FAW while the MON89034 event which has demonstrated efficacy for control of both FAW and stem borers has been cultivated in South Africa since 2010 [101]. MON89034 is recommended for FAW control due to its high efficacy against the pest. The Water Efficient Maize for Africa (WEMA) project now TELA which is operating in 8 countries namely South Africa, Mozambique, Tanzania, Uganda, Kenya, Ethiopia, and Nigeria is to test and deploy the transgenic maize in Africa. Although Genetic engineering would provide an additional intervention that other countries struggling with FAW can explore alongside other integrated pest management practices, there are still contentions on the safety and sustainability of control using Bt maize. Many African countries also lack the legal frameworks for research on and deployment of Genetically Modified Organisms.

6.6 Chemical control of the fall armyworm

Chemical control is one of the key methods for controlling FAW in the Americas and Africa [97]. Initial efforts to control FAW in Africa using chemicals relied on recommendations of effective pesticides from the Americas, leading to the purchase, distribution and application of several synthetic pesticides in many countries. Consequently, many maize farmers have adopted the use of pesticides, a practice that was rare before the advent of FAW. In this section, we review information on the use of chemical pesticides to control the FAW in selected countries on the continent as it is difficult to obtain country-specific information.

A variety of synthetic pesticides have been registered and recommended for control of FAW in different African countries. These include Carbamates, Organophosphates, Ryanodine Receptor modulators, Avermectins, Spinosyns, Oxadiazines, Nereistoxin and Pyrethroids (**Table 2**). Among the above, Pyethroids

Pesticide class	IRAC Group	WHO classification	Active ingredient(s)	Country of registration	Pesticide brand examples
Avermectins	6	II	Abamectin + Emamectin	Uganda[1]	Amdocs
			Abamectin	Malawi[2]	Snowmectin 1.6 EC , Antario
		IIV	Emamectin benzoate	Uganda[1]	Chlobenzo, Prove (EC), Dynamo (WG)
				South Africa[3]	Emma, Proclaim, Promec 20EW, etc.
				Zambia[4]	Prove (EC), Denim Fit 50WG
Benzoylureas	15	II	Lufenuron	South Africa[3]	Judge, Sorba
				Kenya[5]	Heritage 5%, Legacy, Match
			Diflubenzuron	South Africa[3]	Dimilin 25 WP, Dimilin 48 SC
Carbamate	1A	IB	Methomyl	South Africa[3]	Spitfire 900 SP, Cyplamyl 90 SP, Masta 900 SP
		II	Carbosulfan	South Africa[3]	Marshal 48 EC
				Kenya[5]	Marshall 250EC
	14		Cartap hydrochloride	South Africa[3]	Ag-Tap 500 SP
Organophosphates	1B	II	Chlorpyrifos	South Africa[3]	Avi Klorpirifos, Agropyrifos, Pyrinex 480 EC
				Malawi[2]	Chlorpyrifos 480 EC
			Profenofos	Malawi[2]	Snoweron 500 EC
		III	Mercaptothion + Malathion	South Africa[3]	Avi-Merkaptothion DP
			Acephate	Kenya[5]	Lotus 75% SP, Ortran 97, Orthene pellet
Oxadiazine	22A	II	Indoxacarb	South Africa[3]	Doxstar Flo, Advance, Steward, Addition
				Malawi[2]	Steward 150 EC
				Kenya[5]	Merit 150 SC, Avaunt150 SC
				Zambia[4]	Devacarb
				Sudan[6]	Vaulent 150S C

DOI: http://dx.doi.org/10.5772/intechopen.96637

Pesticide class	IRAC Group	WHO classification	Active ingredient(s)	Country of registration	Pesticide brand examples
Pyrethroids	3A	II	Beta- cypermethrin	South Africa[3]	Akito
			Alpha- cypermethrin	Kenya[4]	Bestox 20 EC, Navigator 100 EC
			Deltamethrin	Malawi[2]	Decis Forte, Deltanex 25 EC, Deltmax 25 EC
				Zambia[4]	Decis, Decitab
	28		Lambda-Cyhalothrin	Kenya[4]	Duduthrin Karate
	3A	III	Teflubenzuron+Cypermethrin	Malawi[2]	WormAtak EC
		NL**	Gamma- cyhalothrin	Kenya[5]	Vantex 60 CS
Ryanodine Receptor Modulators	28	III	Flubendiamide	South Africa[3]	Belt 480 SC
				Malawi[2]	Belt 480 SC
				Kenya[5]	Belt 480 SC
		U	Chlorantraniliprole	South Africa[3]	Coragen 20 SC, Prevathon, Mythic FN SC
				Kenya[5]	Coragen 20 SC
Spinosyns	5A	U	Spinetoram	South Africa[3]	Delegate 250 WG
				Kenya[5]	Radiant 120 SC
Nereistoxin analogue	14	U	Cartap hydrochloride	South Africa[3]	Ag-Tap 500
Combinations of pesticide classes					
Avermectin+Diamide	6 and 28	II	Abamectin+Chlorantraniliprole	Kenya[5]	Voliam Targo 063
Benzoylureas+Avermectin	15 and 6	III	Lufenuron+Emamectin benzoate	South Africa[3]	Denim Fit
				Malawi[2]	Proclaim Fit
				Sudan[6]	Denim Fit 50%

Pesticide class	IRAC Group	WHO classification	Active ingredient(s)	Country of registration	Pesticide brand examples
Benzoylureas+Oxadiazine	15 and 22A	U	Novaluron+Indoxacarb	South Africa[3]	Plemax
Carbamate+Pyrethroids	1A and 3A	_	Benfuracarb+fenvalerate	South Africa[3]	Oncol Super 220 EC
Organophosphates+Pyrethroids	1B and 3A	II	Profenofos+Cypermethrin	Uganda[1]	Roket, Agro-Cypro, Supa Profenofos, Hitcell
			Pirimiphos methyl+Deltamethrin	Malawi[2]	Ecoterex 0.5 GR
				Zimbabwe[7]	Ecoterex 0.5 GR
			Cypermethrin+Chlorpyriphos	South Africa[3]	Cyperfos 500 EC
				Zambia[4]	Cyclone 505 EC
Pyrethroids+Neonicotinoids	3A and 4A	III	Lambda Cyahalothrin+Thiamethoxam	Uganda[1]	Striker, Engeo
Diamide+Neonicotinoid	28 and 4A	_	Chlorantraniliprole+Thiamethoxam	Zambia[4]	Fortenza Duo**A**
				Zimbabwe[7]	
Diamide+pyrethroid	28 and 3A	_	Chlorantraniliprole+Lambda cyhalothrin	South Africa[3]	Ampligo
				Zambia[4]	
Spinosyn+Benzoylureas	5 and 18		Spinetoram+methoxyfenozide	South Africa[3]	Uphold 360 SC
Spinosyns+Diamide	5 and 28	U	Spinotetramat 75+Flubendiamide 100	Sudan[6]	Belt extra 175 OD
Biopesticides					
Microbial disruptors of insect midgut membrane	II	U	*Bacillus thuringiensis*	South Africa[3]	Delfin, Florbac WG
			Beauveria bassiana	South Africa[3]	Eco-Bb

**Superscript numbers are references; [1]Ministry of Agriculture Animal Industry and Fisheries. 2018; [2]Pesticides Control Board. 2015 (Malawi); [3]IRAC, 2019. (2018); [4]Simwinga V, Mudenda M, Mumba S. Pest management decision guide: Green and yellow list. Fall armyworm in maize Zambia 2018. Available: https://www.plantwise.org/FullTextPDF/2017/2017780127. [5]Kenyan Pesticide Control Board, 2021; [6]Omer A.Elnour, Eisa Y.Adam, El Rabei A. Obeid and Doula. S.A Salih. 2019; [7]Evaluation of some insecticides against Fall armyworm, Spodoptera frugiperda, J.E. Smith on maize, Zea mays. Scientific Journal Quarterly University of Bakht Alruda, 28: 1858 –6139.*

Table 2.
Selected chemical pesticides registered for control of the Fall armyworm in Africa.

and Organophosphates are the most used (**Table 2**), followed by Avermectins, perhaps because of their availability and lower prices. Combination of different classes of pesticides are also on the market (**Table 2**). The combinations help increase the effectiveness, target spectrum, and reduce the speed with which pesticide resistance can develop. Pesticides shown in **Table 2** below are approved and recommended by the respective African governments to control FAW. To date, several African countries do not have an official register for approved pesticides for control of the FAW. This may be in part due to the failure to test chemical pesticides for their efficacy. Nonetheless, we provide a review of research efforts towards use of chemical pesticides to control the FAW in Africa.

The rapid adoption of pesticides by farmers in Africa was mostly due to the distribution of free pesticides by the governments. Free distribution proved ineffective in most countries because;

1. Distribution was not matched with adequate training of farmers on proper pesticide usage.

2. Most of the distributed pesticides were not effective as they were never evaluated to determine the suitable application rates for the different agroecologies within the continent [2, 47]. This resulted in indiscriminate spraying by farmers.

3. Farmers' negative perceptions on use of pesticides because they believe the latter are not effective in controlling the FAW [56, 102, 103].

4. Recommendations by governments of cheap chemical pesticides known to have developed resistance to FAW elsewhere. For example, Organophosphates and Pyrethroids-Pyrethrins pesticides to which FAW is reported to have developed resistance in the Americas [19–21]. The above has resulted in indiscriminate application of pesticides by farmers causing fears of likely resistance development in the near future [49, 56]. Although, there are no reports of development of pesticide resistance in FAW in Africa, there is evidence of samples of the pest from Kenya, Malawi and Uganda possessing mutations associated with resistance to organophosphates and carbamates [22, 23].

5. Furthermore, effectiveness of pesticide use in Africa is hindered because many of the countries in Africa lack listings of registered (government approved) pesticides for control of the FAW [97].

7. The future FAW management in Africa

The current invasion of the fall armyworm (FAW) in sub-Saharan Africa (SSA) and other parts of the world threatens food and nutrition security of many nations, and also poses challenges in the attainment of a number sustainable development goals (SDGs) including SDGs: 1 (no poverty), 2 (zero hunger), 3 (good health and wellbeing, 8 (decent work and economic growth) and 12 (protecting life on earth). Without sustainable management of this invasive pest, farmers' efforts in increasing cereal productivity, most especially smallholder farmers in nations that rely on maize as a staple crop, will be futile. The previous sections provided detailed information on the biology and ecology of the FAW, and current management practices for curtailing damage and associated yield losses. Like most invasive insect pests, FAW management across the continent, has relied primarily on the use of pesticides and cultural interventions, but with minimal efforts on an IPM approach. In view of the pest status, threat posed to food and nutrition security, and the grave health and environmental footprints of indiscriminate use of pesticides, the need to formulate effective and sustainable management approaches is dire. This calls for formulation of an effective Pest Management Plan (PMP) at the national and regional levels. The plan should consider International Plant Protection Convention (IPPC) policies, the Pesticides and Toxic Substances Regulations, the Plant Pests and Diseases Regulatory Acts and, Environmental and Social Management Frameworks (ESMF). The components of such a plan may include: i) periodic scouting and review of FAW damage and impact to cereal productivity, ii) identification of interactions between FAW and other cereal insect pests belonging to the same family and consequential impact on productivity, iii) exploration of alternative ways of FAW management, with emphasis on environmentally friendly and socially acceptable approaches, and iv) identification of concerns related to pesticide use and recommendation of measures for enhanced public and occupational health and safety.

7.1 Interventions for sustainable management of fall armyworm in Africa

In planning for effective FAW management in the future, the following should be considered at national or regional levels;

7.1.1 Research

Sufficient knowledge or data is a prerequisite for developing any sound action plan and management decision. According to FAO (2016), institutional capacity, effective policies and an enabling environment are key ingredients for the transformation and growth of the agricultural sector in the face of climate change. A lot of effort has been put in understanding the ecology of FAW and possible management practices, but much remains to be done. Equally important is the need to understand FAW behavior in the face of climate change. Thus, future FAW management will require generation of data that can be used for developing predictive models for determining future hotspots/outbreaks and decision-making. Addressing issues of pesticide resistance, will require proper planning, implementation of research activities, monitoring and evaluation, safeguards compliance, and regional engagement.

7.1.2 Collation and dissemination of research outputs

In addition to research, future FAW management should focus on effective dissemination of proven technologies, while leveraging on existing dissemination systems such as extension services of the Ministries of Agriculture and mass communication channels. Regional and/or continental exchange of information, knowledge and technologies will be key to managing this transboundary pest. Currently, however, the national dissemination systems in most African countries are not only weakened by poor research-extension linkages, but by low human capacity; lack of information and communication materials on recommended technologies; lack of harmonization of information packaging; inappropriate packaging of extension messages; limited use of mass communication channels; and inadequate training. Incorporating unique and modern approaches to transfer knowledge across the value chains, e.g., using training of trainers (TOT) approach, use of videos on social media platforms, radios, promoting strategic dialog and/or effective communication for knowledge transfer will play a critical role in sustainable management of FAW.

7.1.3 Capacity building

Currently, most scientists, extension officers and farmers lack knowledge on pest identification, economic impact assessment and management strategies. Capacity building interventions through scientific short and/or long-term trainings, as well as upgrade of research infrastructure (e.g., functional laboratories, greenhouses, containment facilities that can handle GMO materials) and information technology and knowledge management system are required to address the knowledge gaps for sustainable management of FAW on the continent. At grass-root level, hands-on trainings on FAW scouting, identification, safe use of pesticides and ecologically based integrated pest management strategies should be implemented through farmers' field schools. Furthermore, government bodies like the Environmental Management Agencies, Agrochemical Associations, etc.) in different countries should take an active role in building the capacity of extension officers and farmers on good pesticide use and putting in place an effective monitoring system to ensure that only registered pesticides are being distributed to farmers, and that the formulations are maintained in their registered state (without re-formulation).

7.1.4 Enhanced coordination of stakeholders

Across the continent, the response to FAW outbreak by key actors (e.g., national governments, research scientists and regional bodies like FAO) was loosely coordinated. Moving forwards, regional and national government leaders should focus on ensuring that all essential FAW task forces and management coordination functions are effectively carried out.

7.1.5 Effective policy implementation

In most sub-Saharan Africa (SSA) countries, agricultural policies put emphasis on agricultural productivity, with the promotion of pesticide use to address the perennial low productivity placed among top pest management intervention strategies. Legislations on management of Plant Pests and Diseases that is often enforced by the Phyto-sanitary Services Department under the Ministry of Agriculture and The Pesticides and Toxic Substances Regulations (PTSR) do exist across the continent. Unfortunately, issues of improper and unsafe use of chemicals due to inadequate enforcement of regulations and lack of compliance to safety measures pose a serious threat to the ecosystem. Therefore, through the participation of key players (i.e., right institutional structures, systems and set of skilled personnel), the implementation of effective and targeted regulatory policies for improving agricultural sector, e.g., those that support subsidies, grants and tax credits, risk mitigation, market access, purchase of surplus produce from farmers must be strengthened, and applied in future FAW management plans.

While the African Union, through her Comprehensive African Agricultural Development Program has formulated strategies to improve agricultural productivity at regional level, FAW management plans are currently embedded within national agricultural and food security strategic plans but not at regional level. Thus, attempts to transform the African agriculture sector should focus on preparedness to respond to pest threats, specifically invasive transboundary pests like FAW and migratory locusts that the continent has witnessed recently. Such a preparedness plan should include, but not be limited to i) investments in proper policies, pest and disease management, extension and infrastructure, and ii) effective coordination of available resources for pest and disease management.

Addressing the FAW constraint in sub-Saharan Africa requires policy dialog and policy harmonization on key areas that affect the management of the pest at national and regional levels. For instance, issues of IPs/patents for new technologies, harmonization and operationalization of the pesticide regulatory system, implementation of biosafety regulations, etc. should be addressed at the regional level. Approved pesticides for FAW control at regional and national levels must comply with guidelines of the International Plant Protection Convention (IPPC) on Phytosanitary Measures for pest surveillance, risk identification, reporting and management and complement the recently published IPPC Guide to Pest Risk Communication. Equally such harmonized policies should support other global voices, such as The Stockholm and the Rotterdam Conventions that advocate for use of non-Persistent Organic Compounds (POPs), the Pesticide Action Network (PAN) and the Sustainable Agriculture Network (SAN). For example, the Pesticide and Toxic Substance Regulations, which advocates for using pesticides with: i) negligible adverse effects on humans and domestic animals in the treated areas; ii) effectiveness against the target species; iii) minimal effects on non-target species, especially damage to natural enemies, and the environment in general; iv) avoidance of pesticide resistance and resurgence, should be borne in mind during such policy dialogs and harmonization.

Author details

Michael Hilary Otim[1*], Komi Kouma Mokpokpo Fiaboe[2], Juliet Akello[3], Barnabas Mudde[1], Allan Tekkara Obonyom[1], Anani Yaovi Bruce[4], Winnifred Aool Opio[1], Peter Chinwada[3], Girma Hailu[5] and Pamela Paparu[1]

1 National Agricultural Research Organization, Uganda

2 International Institute of Tropical Agriculture, Cameroon

3 International Institute of Tropical Agriculture, Zambia

4 International Maize and Wheat Improvement Center, Kenya

5 International Centre of Insect Physiology and Ecology, Kenya

*Address all correspondence to: motim9405@gmail.com

References

[1] Goergen G, Kumar PL, Sankung SB, Togola A, Tamò M. First report of outbreaks of the fall armyworm *Spodoptera frugiperda* (J E Smith) (Lepidoptera, Noctuidae), a new alien invasive pest in West and Central Africa. PLoS One. 2016;11(10):1-9.

[2] Day R, Abrahams P, Bateman M, Beale T, Clottey V, Cock M, et al. Fall armyworm: Impacts and implications for Africa. Outlooks Pest Manag. 2017;28(5):196-201.

[3] Donna B, Divyanshi W. Year in Review: 2018 in 14 Charts. 2018.

[4] Kassie M, Wossen T, De Groote H, Tefera T, Sevgan S, Balew S. Economic impacts of fall armyworm and its management strategies: Evidence from southern Ethiopia. Eur Rev Agric Econ. 2020;47(4):1473-501.

[5] ten Berge HFM, Hijbeek R, van Loon MP, Rurinda J, Tesfaye K, Zingore S, et al. Maize crop nutrient input requirements for food security in sub-Saharan Africa. Glob Food Sec. 2019 Dec;23:9-21.

[6] Cock MJW, Beseh PK, Buddie AG, Cafá G, Crozier J. Molecular methods to detect *Spodoptera frugiperda* in Ghana, and implications for monitoring the spread of invasive species in developing countries. Sci Rep [Internet]. 2017;7(1):4103. Available from: http://www.nature.com/articles/s41598-017-04238-y

[7] FAO. Briefing Note on FAO actions on fall armyworm [Internet]. Briefing Note 03. 2018. p. 1-6. Available from: http://www.fao.org/fall-armyworm/en/

[8] Nagoshi RN, Koffi D, Agboka K, Tounou KA, Banerjee R, Jurat-Fuentes JL, et al. Comparative molecular analyses of invasive fall armyworm in Togo reveal strong similarities to populations from the eastern United States and the Greater Antilles. PLoS One [Internet]. 2017;12(7):1-15. Available from: http://dx.doi.org/10.1371/journal.pone.0181982

[9] Otim MH, Tay WT, Walsh TK, Kanyesigye D, Adumo S, Abongosi J, et al. Detection of sister-species in invasive populations of the fall armyworm *Spodoptera frugiperda* (Lepidoptera: Noctuidae) from Uganda. PLoS One [Internet]. 2018;13(4):1-18. Available from: http://dx.doi.org/10.1371/journal.pone.0194571

[10] Tindo M, Tagne A, Tigui A, Kengni F, Atanga J, Bila S, et al. First report of the fall army worm, *Spodoptera frugiperda* (Lepidoptera, Noctuidae) in Cameroon. J Biol Biochem Sci [Internet]. 2017;25(March):30-2. Available from: https://www.ippc.int/static/media/files/pestreport/2017/06/02/first_report_Fall_Army_Worm___Cameroon.pdf

[11] Uzayisenga B, Waweru B, Kajuga J, Karangwa P, Uwumukiza B, Edgington S, et al. First record of the fall armyworm, *Spodoptera frugiperda* (J.E. Smith, 1797) (Lepidoptera: Noctuidae), in Rwanda. African Entomol. 2018;26(1):244-6.

[12] Johnson SJ. Migration and the life history strategy of the fall armyworm, *Spodoptera frugiperda* in the western hemisphere. Int J Trop Insect Sci [Internet]. 1987;8(4-5-6):543-9. Available from: https://doi.org/10.1017/S1742758400022591

[13] FAO and CABI. Community-based fall armyworm monitoring, early warning and management: Training of Trainers Manual [Internet]. 2019. 112 p. Available from: http://www.fao.org/3/ca2924en/CA2924EN.pdf

[14] Capinera J. Fall armyworm, *Spodoptera frugiperda* (J.E. Smith) (Insecta: Lepidoptera: Noctuidae) [Internet]. 1999. p. 9. Available from: http://entnemdept.ufl.edu/creatures/field/fall_armyworm.htm

[15] Assefa F, Ayalew D. Status and control measures of fall armyworm (*Spodoptera frugiperda*) infestations in maize fields in Ethiopia: A review. Cogent Food Agric [Internet]. 2019;5(1):1-16. Available from: https://doi.org/10.1080/23311932.2019.1641902

[16] Nagoshi RN, Goergen G, Tounou KA, Agboka K, Koffi D, Meagher RL. Analysis of strain distribution, migratory potential, and invasion history of fall armyworm populations in northern Sub-Saharan Africa. Sci Rep [Internet]. 2018;8(1):1-10. Available from: http://dx.doi.org/10.1038/s41598-018-21954-1

[17] Tay W, Rane R, Padovan A, Walsh T, Elfekih S, Downes S, et al. Global FAW population genomic signature supports complex introduction events across the Old World. bioRxiv [Internet]. 2020;2020.06.12.147660. Available from: http://biorxiv.org/content/early/2020/06/15/2020.06.12.147660.abstract

[18] Pashley DP. Current status of fall armyworm host strains. Florida Entomol. 1988;71(3):227.

[19] Yu SJ. Insecticide resistance in the fall armyworm, *Spodoptera frugiperda* (J. E. Smith). Pestic Biochem Physiol. 1991;39(1):84-91.

[20] Carvalho RA, Omoto C, Field LM, Williamson MS, Bass C. Investigating the molecular mechanisms of organophosphate and pyrethroid resistance in the fall armyworm *Spodoptera frugiperda*. PLoS One. 2013;8(4).

[21] Gutirrez-Moreno R, Mota-Sanchez D, Blanco CA, Whalon ME, Terán-Santofimio H, Rodriguez-Maciel JC, et al. Field-evolved resistance of the fall armyworm (Lepidoptera: Noctuidae) to synthetic insecticides in Puerto Rico and Mexico. J Econ Entomol. 2019;112(2):792-802.

[22] Boaventura D, Martin M, Pozzebon A, Mota-Sanchez D, Nauen R. Monitoring of target-site mutations conferring insecticide resistance in *Spodoptera frugiperda*. Insects. 2020;11(8):1-15.

[23] Guan F, Zhang J, Shen H, Wang X, Padovan A, Walsh TK, et al. Whole-genome sequencing to detect mutations associated with resistance to insecticides and Bt proteins in *Spodoptera frugiperda*. Insect Sci. 2020;1-12.

[24] Montezano DG, Specht A, Sosa-Gómez DR, Roque-Specht VF, Sousa-Silva JC, Paula-Moraes S V., et al. Host plants of *Spodoptera frugiperda* (Lepidoptera: Noctuidae) in the Americas. African Entomol. 2018;26(2):286-300.

[25] Casmuz A, Casmuz A, Juárez ML, Socías MG, Murúa MG, Prieto S, et al. Revisión de los hospederos del gusano cogollero del maíz, *Spodoptera frugiperda* (Lepidoptera: Noctuidae). Rev la Soc Entomológica Argentina. 2010 Dec;

[26] Prasanna B, Huesing JE, Eddy R, Peschke VM. Fall armyworm in Africa: A gudie for Integrated Pest Management. USAID; 2018.

[27] Jeger M, Bragard C, Caffier D, Candresse T, Chatzivassiliou E, Dehnen-Schmutz K, et al. Pest categorisation of *Spodoptera frugiperda*. EFSA J [Internet]. 2017;15(7). Available from: http://doi.wiley.com/10.2903/j.efsa.2017.4927

[28] Bajracharya ASR, Bhat B, Sharma P, Shashank PR, Meshram NM, Hashmi TR. First record of fall

armyworm *Spodoptera frugiperda* (J. E. Smith) from Nepal. Indian J Entomol. 2019;81(4).

[29] Montezano ADG, Specht A, Montezano DG, Specht A. Host Plants of *Spodoptera frugiperda* (Lepidoptera : Noctuidae) in the Americas Published By : Entomological Society of Southern Africa Review article Host plants of *Spodoptera frugiperda* (Lepidoptera : Noctuidae) in the Americas. African Entomol. 2018;26(2):286-300.

[30] Buntin GD. A Review of plant response to fall armyworm, *Spodoptera frugiperda* (J. E. Smith), injury in selected field and forage crops. Florida Entomol. 1986;69(3):549.

[31] Dowd PF. Insect management to facilitate preharvest mycotoxin management. J Toxicol - Toxin Rev. 2003;22(2-3):327-50.

[32] Midingoyi SG, Kassie M, Muriithi B, Diiro G, Ekesi S. Do Farmers and the Environment Benefit from Adopting Integrated Pest Management Practices? Evidence from Kenya. J Agric Econ. 2019 Jun;70(2):452-70.

[33] Adeye AT, Sikirou R, Boukari S, Aboudou M, Amagnide GYGA, Idrissou BS, et al. Protection de la culture de maïs contre *Spodoptera frugiperda* avec les insecticides Plantneem, Lambdace 25 EC et Viper 46 EC et reduction de pertes de rendement au Benin. J la Rech Sci l'Université Lomé. 2018;20(2):53-65.

[34] Houngbo S, Zannou A, Aoudji A, Sossou HC, Sinzogan A, Sikirou R, et al. Farmers' knowledge and management practices of fall armyworm, *Spodoptera frugiperda* (J.E. Smith) in Benin, West Africa. Agric [Internet]. 2020;10(10):1-15. Available from: www.mdpi.com/journal/agriculture

[35] Rwomushana, I. ; Bateman, M. ; Beale, T. ; Beseh, P. ; Cameron, K. ; Chiluba, M. ; Clottey, V. ; Davis, T. ; Day, R. ; Early, R. ; Godwin, J. ; Gonzalez-Moreno, P. ; Kansiime, M. ; Kenis, M. ; Makale, F. ; Mugambi, I. ; Murphy, S. ; Nunda, W. ; Phiri, N. J. Fall armyworm: impacts and implications for Africa [Internet]. Wallingford, UK: CABI International; 2018. Available from: https://www.invasive-species.org/wp-c...

[36] Britz C. Relationship between *Spodoptera frugiperda* (Lepidoptera: Noctuidae) damage and yield loss in maize. 2020.

[37] FAO. Progress report on the implementation of the Global Action. Rome, Italy; 2020.

[38] Phil Abrahams., Melanie Bateman., Tim Beale., Victor Clottey., Matthew Cock., Yelitza Colmenarez., Natalia Corniani., Roger Day., Regan Early., Julien Godwin., Jose Gomez., Pablo Gonzalez Moreno., Sean T. Murphy., Birgitta Oppong-Mensah., Noah Phiri., Cor AW. Fall armyworm: Impacts and implications for Africa [Internet]. Vol. 2, Evidence Note (2). 2017. p. 144. Available from: https://www.invasive-species.org/Uploads/InvasiveSpecies/FAW Evidence Note October 2018.pdf

[39] Kansiime MK, Mugambi I, Rwomushana I, Nunda W, Lamontagne-Godwin J, Rware H, et al. Farmer perception of fall armyworm (*Spodoptera frugiperda* J.E. Smith) and farm-level management practices in Zambia. Pest Manag Sci. 2019;75(10):2840-50.

[40] McGrath, D., Huesing, J.E., Beiriger, R., Nuessly, G., Tepa-Yotto, T.G., Hodson, D., Kimathi, E., Elias, F., Obaje, J.A., Mulaa, M., Paula, A., Mabrouk, A.F.A., and Belayneh Y. Monitoring, surveillance, and scouting for fall armyworm. In: Prasanna, B.M., Huesing, J.E., Eddy, R. and Peschke V., editor. Fall Armyworm in

Africa: A Guide for Integrated Pest Management. First. Mexico: CIMMYT; 2018. p. 11-28.

[41] Hruska AJ. Fall armyworm (*Spodoptera frugiperda*) management by smallholders. CAB Rev Perspect Agric Vet Sci Nutr Nat Resour. 2019;14(043):0-3.

[42] Silvain JF. Use of pheromone traps as a warning system against attacks of *Spodoptera frugiperda* larvae in French Guiana. Florida Entomol. 1986;69(1):139.

[43] Osborne JL, Loxdale HD, Woiwod IP. Monitoring insect dispersal: methods and approaches. In: In: Bullock, J.M., Kenward, R.E. and Hails RS, editor. Dispersal ecology (ed JM Bullock, RE Kenward & RS Hails). Blackwell Publishing, Oxford; 2002. p. 24-49.

[44] Davis F, Ng S, Williams W. Visual rating scales for screening whorl-stage corn for resistance to fall armyworm. In: Mississippi Agricultural and Forestry Experiment Station Technical Bulletin [Internet]. Mississippi State University, MS39762, USA; 1992. p. 1-9. Available from: http://agris.fao.org/agris-search/search/display.do?f=1994/US/US94179.xml;US9406363

[45] Nboyine JA, Kusi F, Abudulai M, Badii BK, Zakaria M, Adu GB, et al. A new pest, *Spodoptera frugiperda* (J.E. Smith), in tropical Africa: Its seasonal dynamics and damage in maize fields in northern Ghana. Crop Prot [Internet]. 2020;127(May 2019):104960. Available from: https://doi.org/10.1016/j.cropro.2019.104960

[46] Sisay B, Simiyu J, Mendesil E, Likhayo P, Ayalew G, Mohamed S, et al. Fall armyworm, *Spodoptera frugiperda* infestations in East Africa: Assessment of damage and parasitism. Insects. 2019;10(7):1-10.

[47] Sisay B, Tefera T, Wakgari M, Ayalew G, Mendesil E. The efficacy of selected synthetic insecticides and botanicals against fall armyworm, *Spodoptera frugiperda*, in maize. Insects. 2019;10(2).

[48] Burtet, L.M., Bernardi, O., Melo, A.A., Pes, M.P., Strahl, T.T., Guedes, J.V., 2017. Managing fall armyworm, *Spodoptera frugiperda* (Lepidoptera: Noctuidae), with Bt maize and insecticides in southern Brazil. Pest Manag. Sci. 73, 2569-2577. https://doi.org/10.1002/ps.4660

[49] Baudron F, Zaman-Allah MA, Chaipa I, Chari N, Chinwada P. Understanding the factors influencing fall armyworm (*Spodoptera frugiperda* J.E. Smith) damage in African smallholder maize fields and quantifying its impact on yield. A case study in Eastern Zimbabwe. Crop Prot [Internet]. 2019 Jun;120:141-50. Available from: https://doi.org/10.1016/j.cropro.2019.01.028

[50] Rodríguez-del-Bosque LA, Cantú-Almaguer MA, Reyes-Méndez CA. Corn hybrids and planting dates affect yield losses by *Helicoverpa zea* and *Spodoptera frugiperda* (Lepidoptera: Noctuidae) feeding on ears in Mexico. J Entomol Sci [Internet]. 2012 Apr;47(2):177-84. Available from: http://www.bioone.org/doi/10.18474/0749-8004-47.2.177

[51] Wezel A, Bellon S, Doré T, Francis C, Vallod D, David C. Agro-ecology as a science, a movement and a practice. Sustain Agric. 2009;2:27-43.

[52] Altieri MA, Nicholls CI, Montalba R. Technological approaches to sustainable agriculture at a crossroads: An agroecological perspective. Sustainability. 2017;9:1-13.

[53] FAO (Food and Agriculture Organization of the United Nations).

Integrated management of the fall armyworm on maize [Internet]. 2018. 1-133 p. Available from: http://www.grainsa.co.za/upload/FAO---FAW-Guide.pdf

[54] Hailu G, Niassy S, Zeyaur KR, Ochatum N, Subramanian S. Maize–legume intercropping and push–pull for management of fall armyworm, stemborers, and striga in Uganda. Agron J. 2018;110(6):2513-22.

[55] Midega CAO, Pittchar JO, Pickett JA, Hailu GW, Khan ZR. A climate-adapted push-pull system effectively controls fall armyworm, *Spodoptera frugiperda* (J E Smith), in maize in East Africa. Crop Prot [Internet]. 2018;105(November 2017):10-5. Available from: http://linkinghub.elsevier.com/retrieve/pii/S0261219417303216

[56] Kumela T, Simiyu J, Sisay B, Likhayo P, Mendesil E, Gohole L, et al. Farmers' knowledge, perceptions, and management practices of the new invasive pest, fall armyworm (*Spodoptera frugiperda*) in Ethiopia and Kenya. Int J Pest Manag [Internet]. 2019 Jan 2;65(1):1-9. Available from: https://doi.org/10.1080/09670874.2017.1423129

[57] Abate, T., van Huis, A. and AMpofo JK. Pest management strategies in traditional agriculture. Annu Rev Entomol. 2000;45:631-59.

[58] Babendreier, Dirk; Agboyi,; Lakpo Koku; Beseh, Patrick; Osae, Michael; Nboyine, Jerry; Ofori, Selorm E K; Frimpong, Justice O; Clottey, Victor Attuquaye; Kenis M. The efficacy of alternative, environmentally friendly plant protection measures for control of fall armyworm, *Spodoptera frugiperda*, in maize. Insects. 2020;11(January 2016).

[59] Luginbill P. The fall armyworm. Vol. 34, Technical Bulletin. 1928. p. 1-92.

[60] Harrison RD, Thierfelder C, Baudron F, Chinwada P, Midega C, Schaffner U, et al. Agro-ecological options for fall armyworm (*Spodoptera frugiperda* JE Smith)management: Providing low-cost, smallholder friendly solutions to an invasive pest. J Environ Manage. 2019;243(March):318-30.

[61] Oerke EC, Dehne HW. Safeguarding production - Losses in major crops and the role of crop protection. Crop Prot. 2004;23(4):275-85.

[62] Matusso JMM, Mucheru-Muna. Potential role of cereal-legume intercropping systems in integrated soil fertility management in smallholder farming systems of sub-Saharan Africa. Res J Agric Environ Manag [Internet]. 2014;3(3):162-74. Available from: http://www.apexjournal.org

[63] Ngwira AR, Aune JB, Mkwinda S. On-farm evaluation of yield and economic benefit of short term maize legume intercropping systems under conservation agriculture in Malawi. F Crop Res [Internet]. 2012;132:149-57. Available from: http://dx.doi.org/10.1016/j.fcr.2011.12.014

[64] Altieri MA, Letourneau DK, Risch SJ. Vegetation diversity and insect pest outbreaks. CRC Crit Rev Plant Sci [Internet]. 1984 Jan 2;2(2):131-69. Available from: http://www.tandfonline.com/doi/abs/10.1080/07352688409382193

[65] Trenbath BR. Intercropping for the management of pests and diseases. F Crop Res. 1993;34(3-4):381-405.

[66] Tanyi CB, Tanyi CB, Nkongho RN, Okolle JN, Tening AS, Ngosong C. Effect of intercropping beans with maize and botanical extract on fall armyworm (*Spodoptera frugiperda*) Infestation. Int J Agron. 2020;2020.

[67] Deshmukh SS, Prasanna BM, Kalleshwaraswamy CM, Jaba J,

Choudhary B. Fall armyworm (*Spodoptera frugiperda*). In: Omkar, editor. Polyphagous Pests of Crops [Internet]. Singapore: Springer Singapore; 2021. p. 349-72. Available from: http://link.springer.com/10.1007/978-981-15-8075-8_8

[68] Fanadzo M, Dalicuba M, Dube E. Application of conservation agriculture principles for the management of field crops pests. Sustain Agric Rev [Internet]. 2018;125-52. Available from: http://edis.ifas.ufl.edu/in255

[69] All JN. Fall armyworm (Lepidoptera: Noctuidae) infestations in no-tillage cropping systems. Florida Entomol. 1988;71(3):268.

[70] Hassanali A, Herren H, Khan ZR, Pickett JA, Woodcock CM. Integrated pest management: The push-pull approach for controlling insect pests and weeds of cereals, and its potential for other agricultural systems including animal husbandry. Philos Trans R Soc B Biol Sci. 2008;363(1491):611-21.

[71] Caniço A, Mexia A, Santos L. Seasonal dynamics of the alien invasive insect pest *Spodoptera frugiperda* smith (Lepidoptera: Noctuidae) in manica province, central Mozambique. Insects. 2020;11(8):1-12.

[72] Amadou L, Baoua I, Malick NB, Karimoune L, Muniappan R. Native parasitoids recruited by the invasive fall armyworm in Niger. Indian J Entomol [Internet]. 2018;80(4):1253-4. Available from: http://www.indianjournals.com/ijor.aspx?target=ijor:ije&volume=80&issue=4&article=004

[73] Sisay B, Simiyu J, Malusi P, Likhayo P, Mendesil E, Elibariki N, et al. First report of the fall armyworm, *Spodoptera frugiperda* (Lepidoptera: Noctuidae), natural enemies from Africa. J Appl Entomol. 2018;142(8):800-4.

[74] Kenis M, du Plessis H, Van den Berg J, Ba MN, Goergen G, Kwadjo KE, et al. *Telenomus remus*, a candidate parasitoid for the biological control of *Spodoptera frugiperda* in Africa, is already present on the continent. Insects. 2019;10(4):1-10.

[75] Agboyi LK, Goergen G, Beseh P, Mensah SA, Clottey VA, Glikpo R, et al. Parasitoid complex of fall armyworm, *Spodoptera frugiperda*, in Ghana and Benin. Insects. 2020;11(2):1-15.

[76] Caniço A, Mexia A, Santos L. First report of native parasitoids of fall armyworm *Spodoptera frugiperda* smith (Lepidoptera: Noctuidae) in Mozambique. Insects. 2020;11(9):1-12.

[77] Koffi D, Kyerematen R, Eziah VY, Agboka K, Adom M, Goergen G, et al. Natural enemies of the fall armyworm, *Spodoptera frugiperda* (J.E. Smith) (Lepidoptera: Noctuidae) in Ghana. Florida Entomol. 2020 Apr 8;103(1):85.

[78] Otim MH, Aropet SA, Opio M, Kanyesigye D, Opolot HN. Parasitoid distribution and parasitism of the fall armyworm *Spodoptera frugiperda* (Lepidoptera : Noctuidae) in different maize producing regions of Uganda. Insects. 2021;1-20.

[79] Abang AF, Fotso Kuate A, Nanga Nanga S, Okomo Esi RM, Ndemah R, Masso C, et al. Spatio-temporal partitioning and sharing of parasitoids by fall armyworm and maize stemborers in Cameroon. J Appl Entomol. 2020;(July):1-10.

[80] Molina-Ochoa J, Carpenter JE, Heinrichs EA, Foster JE. Parasitoids and parasites of *Spodoptera frugiperda* (Lepidoptera: Noctuidae) in the Americas and Caribbean Basin: An inventory. Florida Entomol. 2003;86(3):254-89.

[81] Kittel RN, Austin AD, Klopfstein S. Molecular and

morphological phylogenetics of chelonine parasitoid wasps (Hymenoptera: Braconidae), with a critical assessment of divergence time estimations. Mol Phylogenet Evol [Internet]. 2016;101:224-41. Available from: http://www.sciencedirect.com/science/article/pii/S1055790316300975

[82] Laminou SA, Ba MN, Karimoune L, Doumma A, Muniappan R. Parasitism of locally recruited egg parasitoids of the fall armyworm in Africa. Insects. 2020;11(7):1-13.

[83] Fiaboe KR. Evaluation of larvicidal potency of some bio-insecticides for the management of Fall Armyworm, *Spodoptera frugiperda* (J.E. Smith) (Lepidoptera: Noctuidae) and inventory of its associated natural enemies. MPhil thesis, University of Ghana, Legon.; 2019.

[84] Akutse KS, Subramanian S, Maniania NK, Dubois T, Ekesi S. Biopesticide research and product development in Africa for sustainable agriculture and food security – Experiences from the International Centre of Insect Physiology and Ecology (icipe). Front Sustain Food Syst. 2020;4(September):1-14.

[85] Fiaboe K. Evaluation of larvicidal potency of some bio-insecticides for the management of fall armyworm, *Spodoptera frugiperda* (J.E. Smith) (Lepidoptera: Noctuidae) and inventory of its associated natural enemies. University of Ghana; 2019.

[86] Ngangambe MH, Mwatawala MW. Effects of entomopathogenic fungi (EPFs) and cropping systems on parasitoids of fall armyworm (*Spodoptera frugiperda*) on maize in eastern central, Tanzania. Biocontrol Sci Technol [Internet]. 2020;30(5):418-30. Available from: https://doi.org/10.1080/09583157.2020.1726878

[87] Waiss AC, Chan BG, Elliger CA. Host plant resistance to insects. In: Host Plant Resistance to Pests [Internet]. American Chemical Society; 1977. p. 115-128 SE – 8. (ACS Symposium Series; vol. 62). Available from: https://doi.org/10.1021/bk-1977-0062.ch008

[88] Sharma HC, Ortiz R. Host plant resistance to insects: An eco-friendly approach for pest management and environment conservation. J Environ Biol. 2002;23(2):111-35.

[89] Mihm JA, Smith ME, Deutsch JA. Development of Open-Pollinated Varieties, Non-Conventional Hybrids and Inbred Lines of Tropical Maize with Resistance to Fall armyworm, *Spodoptera frugiperda* (Lepidoptera: Noctuidae), at CIMMYT. Florida Entomol. 1988;71(3):262.

[90] Scott GE, Davis FM, Williams WP. Registration of MP701 and MP702 Germplasm Lines of Maize1 (Reg. No. GP119 and GP120). Crop Sci [Internet]. 1982;22(6):cropsci1982.0011183X002200060070x. Available from: https://acsess.onlinelibrary.wiley.com/doi/abs/10.2135/cropsci1982.0011183X002200060070x

[91] Warren HL. Registration H110 and H111 maize germplasm1 (Reg. No. GP 121 and GP 122). Crop Sci [Internet]. 1982;22(6):cropsci1982.0011183X002200060071x. Available from: https://acsess.onlinelibrary.wiley.com/doi/abs/10.2135/cropsci1982.0011183X002200060071x

[92] Williams WP, Davis FM, Windham GL. Registration of Mp708 germplasm Line of Maize. Crop Sci [Internet]. 1990;30(3):cropsci1990.0011183X003000030082x. Available from: https://acsess.onlinelibrary.wiley.com/doi/abs/10.2135/cropsci1990.0011183X003000030082x

[93] Williams WP, Davis FM. Registration of Maize Germplasm Line

Mp716. Crop Sci [Internet]. 2002 Mar 1;42(2):671-2. Available from: https://doi.org/10.2135/cropsci2002.671a

[94] Williams WP, Davis FM. Registration of maize germplasms Mp713 and Mp714. Crop Sci [Internet]. 2000 Mar 1;40(2):584. Available from: https://doi.org/10.2135/cropsci2000.0015rgp

[95] Williams WP, Davis FM. Registration of Mp704 mermplasm Line of Maize1 (Reg. No. GP116). Crop Sci [Internet]. 1982 Nov 1;22(6):cropsci1982.0011183X002200060068x. Available from: https://doi.org/10.2135/cropsci1982.0011183X002200060068x

[96] CIMMYT. A complete listing of maize germplasm from CIMMYT. In: Maize Program Special Report. CIMMYT, Mexico, DF; 1998.

[97] Prasanna B, Huesing JE, Eddy R, Peschke VM. Fall armyworm in Africa: a guide for Integrated Pest Management [Internet]. First. Prasanna. B.M., Joseph E. Huesing, Regina Eddy VMP, editor. Vol. First Edit, Mexico, CDMX: CIMMYT. Mexico; 2018. 45-62 p. Available from: www.maize.org.

[98] CIMMYT. Announcing CIMMYT derived fall armyworm tolerant elite maize hybrids for eastern and southern Africa. CIMMYT, Mexico, DF; 2020.

[99] Jouanin L, Bonadé-Bottino M, Girard C, Morrot G, Giband M. Transgenic plants for insect resistance. Plant Sci. 1998;131(1):1-11.

[100] Niu Y, Meagher RL, Yang F, Huang F. Susceptibility of field populations of the fall armyworm (Lepidoptera: Noctuidae) from Florida and Puerto Rico to Purified Cry1f protein and corn leaf tissue containing single and pyramided Bt Genes. Florida Entomol. 2013;96(3):701-13.

[101] Botha AS, Erasmus A, du Plessis H, Van Den Berg J. Efficacy of Bt Maize for Control of *Spodoptera frugiperda* (Lepidoptera: Noctuidae) in South Africa. J Econ Entomol [Internet]. 2019 May 22;112(3):1260-6. Available from: https://academic.oup.com/jee/advance-article/doi/10.1093/jee/toz048/5378670

[102] Ahmad M, Iqbal Arif M, Ahmad M. Occurrence of insecticide resistance in field populations of *Spodoptera litura* (Lepidoptera: Noctuidae) in Pakistan. Crop Prot. 2007;26(6):809-17.

[103] De Groote H, Kimenju SC, Munyua B, Palmas S, Kassie M, Bruce A. Spread and impact of fall armyworm (*Spodoptera frugiperda* J.E. Smith) in maize production areas of Kenya. Agric Ecosyst Environ [Internet]. 2020;292(December 2019):106804. Available from: https://doi.org/10.1016/j.agee.2019.106804

Chapter 2

RNAi-Mediated Control of Lepidopteran Pests of Important Crop Plants

Shipra Saxena, Sneha Yogindran, Manmohan Arya, Yogita Sharma and Chandra Pal Singh

Abstract

Insects as pests destroy annually an estimated 18–20% of the crop production worldwide. Caterpillars, the larval stage of moths, are the major pests of agricultural products owing to their voracious feeding habits. In the past few decades, the potent methods of insect control, such as insecticides and Bt toxins, have been constrained as a result of health hazards, environmental issues, and development of resistance, after their prolonged application. Thus, there is need to find alternative options to improve plant protection strategies. Recently, RNA interference (RNAi), the post-transcriptional gene-silencing mechanism, has emerged as one of such a novel, sustainable, and environment friendly approaches for insect management and crop protection. RNAi technology relies on selection of a vital insect pest target gene and its expression as a double stranded RNA or stem-loop RNA molecule, which is recognized by the host RNAi machinery and processed into small interfering RNAs (siRNAs) or microRNAs (miRNAs). The siRNA/miRNA along with the RNA-induced silencing complex (RISC) binds to the complimentary mRNA and induce gene silencing at post-transcriptional level. With effective target-gene selection and transgenic plants expressing these precursor RNA molecules, insect pests of various crops have been efficiently managed. In this chapter, we discuss the basic mechanism of RNAi and its application in controlling lepidopteran pests of important crop plants.

Keywords: RNAi, lepidopteran pests, crop protection, plant-mediated RNAi, insect resistant transgenics

1. Introduction

The year 2020 has been declared as the International Year of Plant Health by the United Nations General Assembly, to contemplate over the issue of feeding 10 billion people by 2050 and raise global awareness about the challenge modern agriculture is going to face too in a profitable, efficient, and sustainable way. The challenge would be hardened by additional factors like climate change, decrease in arable land due to degradation, and urban expansion, as well as need for more nutritious food [1]. A major hindrance to crop production is loss by insect pests right from the seedling stage to the post harvesting stage of the product. These losses lead to reduced yields, decreased quality, and thus food insecurity resulting in the deaths of millions of people throughout the world and impacting trade and economy of many developing countries. Annually, 20–40% of global crop production is lost due to pests valued more than US$ 70 billion [2]. Moreover, in the coming years with the increasing global temperatures, plant scientists expect a 10 to 25% increase in crop damage due to insect pests, majorly in the temperate regions [3]. In the class Insecta, the order Lepidoptera, represents the second largest order, with 180,000 species in 128 families and 47 superfamilies. Amongst these, more than 160,000 species are moths [4]. Moths are known for their economic values as the silkworm *Bombyx mori*, as well as a food product like larvae of *Gonimbrasia belina* and *Usta Terpsichore* [5]. The larval stage of moths are major pests of agricultural and forest products pests in most parts of the world [6–8].

The most common method of crop protection from insect pests is calendar-based spraying of insecticides. However, these chemicals cause an increased cost of production, residual toxicity, resistance issues, outbreaks of secondary pests, and potential health hazards on humans and environmental threats [9]. Considering these issues, genetic engineering has emerged as an effective way to control the pest population. Use of Bt toxins from soil bacterium *Bacillus thuringiensis,* has shown great potential in controlling the devastating insect pest population. The bacteria produce insecticidal crystal proteins (ICP), such as Cry and/or Cyt proteins called δ-endotoxins that interact with receptors present in the insect midgut cells. This interaction activates the host proteases and results in oligomeric pore formation, which leads to ionic imbalance in the cell ultimately killing the insects [10–12]. Bt based bio-insecticides have been successfully employed against lepidopteran, dipteran and coleopteran larvae [13, 14]. However, topical application does not last long due to the degradation by UV light, weather and certain proteases [15]. This problem was addressed by introducing the Cry genes into the plants through genetic engineering [16–18]. Genetic transformation of plants to express Bt toxins resulted into enhanced tolerance towards the pests and helped the farmers to control the infestation. Apart from Bt proteins, other insecticidal proteins such as vegetative insecticidal proteins, chitinases, α-amylase inhibitors, protease inhibitors etc., have been shown also to control the pest population [19]. The use of Bt toxins and other proteins to generate transgenic crops has been reviewed by [19, 20]. However, various recent studies have demonstrated that insects have gained resistance towards the Bt proteins in the field [21]. Thus, finding alternative options to improve plant protection strategies is critical to secure global food production for the next decades.

In the past few decades, RNA interference (RNAi), a natural defense mechanism by sequence specific down regulation of cognate mRNA, has emerged as a reverse genetics tool for functional genomics along with various practical applications in areas of therapeutics, agriculture etc. RNAi as a technology has shown immense potential in the area of crop improvement traits like introduction of male sterility, enhancement of nutritional contents, reduction of amount of food allergens and toxic compounds, disease and pest resistance, resistance against various abiotic stresses and enhanced production of secondary metabolites. Down-regulation of insect genes through RNAi has been efficiently used to control insect pests in various crop plants [22–24]. The present chapter focuses on the basic RNAi mechanism in insects and the application of this natural defense machinery in controlling the pest population of some important crop plants and widely consumed vegetable crops.

2. Basics of RNAi mechanism

RNAi is a natural phenomenon of gene regulation that occurs at the post-transcriptional level [25]. Though the discovery of RNAi was demonstrated through exogenous delivery of dsRNA against unc22 gene into *Caenorhabditis elegans* [26], now it is clear that RNAi was operational from long back in plants against RNA viruses, known as Virus-induced gene silencing [27]. Most of the higher eukaryotes including animals, plants and insects, possess RNAi mechanism for silencing the genes in a sequence-specific manner [28]. In RNAi-governed gene silencing, the two classes of small non-coding RNAs play key role, which are small interfering RNAs (siRNAs) and microRNAs (miRNAs). SiRNAs and miRNAs are generated from the double-stranded RNA (dsRNA) and stem-loop RNA precursors, respectively [28]. The precursors of miRNAs are abundantly present in cell endogenously, while major sources of dsRNAs are exogenous. [28, 29]. Average size of mature siRNA and miRNA fall in the range of 21–23 nucleotides. The basic process of RNAi consists of several steps and requires the involvement of RNase enzymes and RNA-binding proteins [26–29]. The Dicer, an RNaseIII type endonuclease, plays a crucial role in processing the dsRNA into 20–25 bp long siRNAs in the cytoplasm (**Figure 1**) [27–31]. Whereas the production of mature miRNA duplex requires multiple processing by the Dicer and other co-factors in the nucleus. First primary-miRNA is processed into a single stem-loop bearing structure referred to as precursor-miRNA (pre-miRNA). Subsequently, pre-miRNA is cleaved and mature miRNA duplex is produced. In plants, multiple Dicer enzymes have been identified and distinctively referred to as Dicer-like proteins (DCL) [31]. The produced mature duplexes of siRNA and miRNA contain a two-nucleotide long overhang at both the 3′ ends [28, 32]. Each duplex of siRNA and miRNA initiates the formation of the RNA-induced silencing complex (RISC) [28, 33]. However, mature RISC is a multi-protein complex that possesses Argonaute protein (Ago) as the core effector molecule in both siRNA and miRNA pathways [28, 33]. Upon incorporation into the RISC, siRNA/miRNA duplex loses one of the strands known as the passenger (sense) strand [28, 33]. While the other strand, referred to as the guide strand (antisense), remains loaded on the RISC and further directs the complex to search for the cognate target mRNA. SiRNA/miRNA finds the specific targets based on the complementarity between siRNA/miRNA and mRNA target sequences [28, 29, 33]. In most of the instances in plants, the perfect complementary base-pairing between siRNA/miRNA and mRNA target induces the endonuclease activity of Ago resulting in cleavage of the target

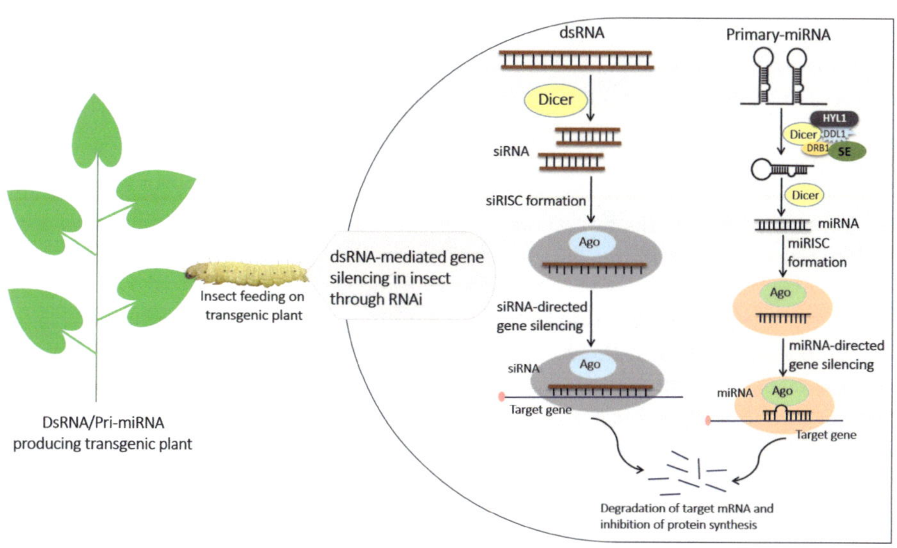

Figure 1.
An overview of dsRNA-mediated knockdown for insect genes through RNAi mechanism in transgenic plants.

mRNA and suppression of the translation [28, 29, 33]. While in insects the mostly miRNAs bind to the cognate mRNA target via partial complementary and leads to the translation repression.

3. RNAi as a pest control technology

The potential of specific gene targeting made RNAi an important method to be applied for plant protection against insect pests [32, 34, 35]. This novel approach provides an opportunity to target any essential gene of the insect. The dsRNA against specific insect gene is expressed from a construct harboring the sense and antisense RNA in the form of DNA. The cloned fragment can be transferred into the plant via agrobacterium-mediated transformation method or in vitro produced dsRNAs/siRNAs can be directly applied to the plants or dsRNA-expressing bacteria are spread on plants as insecticides [36]. Feeding of insects on these dsRNA/siRNAs leads to induction of RNAi-mediated gene silencing (**Figure 1**) [37]. Insects receive either dsRNA or siRNA from these plants wherein siRNAs are straightly incorporated into the RISC, the dsRNA first gets processed into several siRNA molecules in the insect's gut by DCL, then turns to RISC loading [30]. Generated siRNA molecules target the specific insect gene against which they were originally designed based on the sequence complementary between siRNA and target mRNA (**Figure 1**) [32, 34, 35, 37]. This results in suppressed insect growth or mortality [32, 34, 35, 37]. However, miRNA-based gene silencing is achieved through expressing stem-loop bearing primary or precursor miRNA in the plants. Which subsequently get processed by the miRNA-pathway components to give rise mature miRNA duplex and regulates the expression of specific mRNA target.

As a crop protection method, RNAi-based strategies offer the following advantages over the other conventional methods such as chemical insecticides, biological control, or protein-coding transgenes [38]:

1. Highly specific- targets only the intended pest – minimal or null impact on non-target organisms (pollinators, parasitoids, predators and vertebrates)

2. Biodegradable – environment friendly and minimal risk to human health

3. Non-toxic- a natural product as dsRNA is either produced enzymatically in vitro or in vivo through engineered bacteria or host plant

4. No protein production involved

5. It can act individually as well as synergistically with conventional approaches like insecticides and Bt.

Depending upon the method of production of dsRNA and its subsequent delivery to the target pest, there are two major approaches for RNAi-mediated crop protection, topical application of dsRNA (non-transformative) through spraying/injection/root drenching, etc. and generation of transgenic plants expressing dsRNA. Topical application of dsRNA products has been demonstrated through foliar application [39], trunk-injection [39, 40], irrigation [39, 41] and microbe-based [42–45]. Recently, a biotechnology company called RNAagri reported mass making of the encapsulated ready-to-spray dsRNA, APSE RNA Containers (ARCs) by engineered *Escherichia coli* [46]. These non-transformative approaches are advantageous in terms of quick development and testing, no regulation in development of dsRNA as like in a GM product and silencing of genes without introduction of heritable changes into the host plant genome. However, there are several concerns for use of this approach as crop protection method like up-take restrictions, requirement of periodical applications and temporary protection [47].

4. Plant-mediated RNAi for the control of lepidopteran pests

The use of RNAi to silence the insect genes through topical application of dsRNAs has several concerns for use of this approach as crop protection method like up-take restrictions, degradation under field condition thus requirement of periodical applications and temporary protection [47]. Also, the production of large amount of dsRNA is not only expensive, but requires expertise for handling and storage. These limitations can be dealt with a transformative approach of crop protection, generation of transgenic plants expressing double-stranded RNAs (dsRNAs) that target essential genes of insect pests. Feeding upon the plant induces an RNAi response, which either harms, or ideally kills, the pest. Most transgenesis events perform nuclear transformation with agrobacterium vectors carrying inverted repeats of target insect gene sequences. Target gene dsRNA is transcribed by plants RNAi machinery and is processed into siRNAs. However, these plant-processed siRNAs are less efficient in insect cells as compared to longer dsRNA [48]. Another approach to create transgenics is transformation of chloroplast (plastid) DNA. Lack of RNAi machinery in the organelle prevents chopping of dsRNA by Dicer and thereby permits accumulation of much higher amounts of long dsRNA [49].

Popularly known as plant-mediated RNAi or Host-Induced Gene Silencing (HIGS), this strategy has been demonstrated for protection of a range of crops against their specific pest insects, mites, ticks, plant pathogens, viruses, nematodes, and weeds [50–56]. Recently, two regulatory authorities, the Canadian Food Inspection Agency and US Environmental Protection Agency, have declared the approval of the RNAi-based corn event Monsanto MON87411, the "SmartStax PRO" for release and commercialization. The transgenic plants harbor a dsRNA construct that specifically targets the SUCROSE-NON-FERMENTING7 gene of WCR (DvSnf7), together with two insecticidal proteins Cry3Bb1 and Cry34Ab1/Cry35Ab1 [57]. This is in concurrence with the approval granted for apple and potato expressing dsRNAs for quality enhancement [58, 59]. In the past decades, lepidopteran pests have been successfully managed by the first-generation insecticidal plants expressing the Bt proteins. However, with reports of resistance evolution to Bt proteins, scientific community searched for alternatives to manage these pests. The caterpillar pests were one of the first and main targets for RNAi transgenics.

4.1 Model plants

To prove a hypothesis, with the available human and financial resources and carry forward the research as rapidly as possible, researchers use model systems. Model plants like *Arabidopsis thaliana* and *Nicotiana tabacum* can be easily manipulated, are genetically tractable, and about them much is already known. The very first report of plant-mediated RNAi for lepidopteran insect resistance was published by Mao et al. (2007) where they have silenced cytochrome P450 monooxygenase CTP6AE14 gene of *Helicoverpa armigera* involved in degradation of gossypol. The dsRNA was expressed in model plants *A. thaliana* and *N. tabacum* which when fed to insect *H. armigera* resulted into significant reduction in the transcript level, augmented gossypol toxicity in larvae and affected the larval weight and size [60].

The same group also tested the efficiency of plant-mediated RNAi in silencing other midgut gene, GST1, which encodes a glutathione-S-transferase and which is not affected by gossypol content. Feeding of transgenic *A. thaliana* plants resulted into decreased transcript level in the insect midgut and also resulted into larval weight reduction [60]. Insects are compelled to undergo molting through shedding the old cuticles during their growth and enter into pupal stage after which they metamorphose into adult moth. Hence, the whole process of molting is a vital way to regulate development. The major genes associated with molting are the target of insect specific chemical pesticides which have shown promising result [61]. 20, hydroxyecdysone is one of the main genes involved in molting process and metamorphosis and dsRNA expressing tobacco against this gene resulted into impaired molting, pupation and adult emergence rate in *H. armigera* and *Spodoptera exigua* [62]. Silencing of molt-regulating transcription factor, hormone receptor 3 (HR3), of *H. armigera* also resulted into significant downregulation of the target gene which affected the molting and larval growth cycle [63]. Another gene named arginine kinase, required for cellular energy metabolism when silenced through *Arabidopsis*, resulted into defective larval growth and survival in *H. armigera* [64]. Transgenic tobacco plants expressing dsRNA against chitin synthase, cytochrome P450 monooxygenase and V-ATPase genes of *H. armigera* significantly reduced the transcript level and affected the larval weight and pupation [65]. Down-regulation CYP6B46 gene of *Manduca sexta* required for nicotine degradation through genetically modified tobacco resulted into decreased transcripts of the target gene and affected the larval weight [66]. The transgenic plants expressing dsRNA can also be used for the management of closely related insects. *Nicotiana attenuate* plants expressing dsRNA of *M. sexta's* midgut-expressed genes, the nicotine-ingestion induced cytochrome P450 monooxygenase and the lyciumoside-IV-ingestion induced β-glucosidase1, was also able to silence the homologous genes in native *Manduca quinquemaculata*. Hence, careful selection of target genes will help in effective control of congeneric insect pests that share sufficient sequence similarity [67].

4.2 Food and cash crops

4.2.1 Rice

Rice, a staple food for more than half of the global population, is heavily infested by lepidopteran pest, striped stem borer (SSB), *Chilo suppressalis* Walker. The crop yield is significantly reduced by the insect pest as it causes 'deadheart' at the tillering stage and 'whitehead' at the heading stage. In an attempt to impart insect resistance, Jiang & co-workers generated transgenic rice overexpressing five important SSB housekeeping genes, but none of the acquired dsRNA-transgenic rice plants presented significant effects on SSB growth and development. In their subsequent attempt they selected 13 SSB novel microRNAs (miRNAs), and overexpressed them in rice using artificial miRNA (amiRNA) expression technology. Feeding tests on transgenics demonstrated that two out of 13 selected SSB novel miRNAs caused significant growth inhibition in SSB [68]. Recently, Zheng et al., (2020) have developed highly SSB-resistant rice (named csu260) expressing amiRNA of SSB endogenous miRNA - miR260 which negatively regulates ecdysteroid biosynthesis, through amiRNA expression technology [69].

4.2.2 Maize

Even though the only commercialized example of HIGS technology is in maize against western corn rootworm (*Diabrotica virgifera*), the technology has not been yet used against the lepidopteran pests. One of the breakthrough research showing the use of RNAi for insect control was performed in maize. The maize plant was transformed using putative V-ATPaseA coding region from WCR. The F1 hybrid plants displayed resistance to wcr evidenced by less nodal injury and healthy root masses [70]. This provides a sufficient possibility of using this technology for controlling lepidopteran pests as well in the near future.

4.2.3 Soyabean

Soybean (*Glycine max* (L.) Merrill) is an important protein and oil-seed agricultural crop worldwide. *Leguminivora glycinivorella* is a major pest of soybean causing direct loss in yield as well as additional losses in the quality and sale price caused by damaged seeds (Edmonds et al. 2000). Silencing of ribosomal protein P0, involved in protein translation and DNA repair through transgenic plants conferred resistance against the pest. Larval mortality, lesser foliage damage, reduced SpbPo expression and developmental deformities were observed in pest after feeding upon the transgenic plants [71]. In another study transgenic soybean plants expressing Spb18S dsRNA also exhibited resistance to the soybean pod borer. Feeding upon the transgenic plants downregulated Spb18S expression levels as well as reduced second-instar larval survival rates. Also, the developed transgenic plants were less damaged by SPB larvae than control plants under field conditions [72].

4.2.4 Cotton

Cotton is cultivated for its soft fiber immensely used in the clothing industries. However, the production is hugely affected by cotton bollworm (*H. armigera*) which not only affect cotton, but also other crop plants as discussed above. Transgenic cotton plants expressing a P450 monooxygenase gene, CYP6AE14, from *H. armigera* showed enhanced resistance towards the pests suggesting the efficacy of RNAi as a tool for pest management [73].

4.3 Common vegetables

Vegetables provide a reasonable source of vitamins and minerals for maintaining good health and also offer economic revenue to combat rural unemployment. Globally, one billion metric tons of vegetables are harvested per year, with Asia being the leading producer. Tomato, *Solanum lycopersicum*, is an important vegetable crop grown and consumed worldwide. An annual production of about 160 million tonnes is harvested globally. However, the production is hugely affected by the various insect pests like fruit-worms, aphids, cutworms, tomato hornworms, tobacco hornworms, cabbage loopers, whiteflies, flea beetles, red spider mite, slugs, and Colorado potato beetles [19]. Tomato yield loss reported only due to insect infestation accounts to 5–55% [74]. RNAi-mediated crop protection in tomato crop has shown promising results. Most insects cause damage to the plants during their larval stage and hence genes regulating the metamorphosis and development are considered as potent targets for successful RNAi-mediated gene silencing. Silencing of juvenile hormone (JH), a sesquiterpene, has been reported to affect the larval growth and development in tomato fruitworm (*H. armigera*). Juvenile hormone acid O-methyl transferase gene (JHAMT), a key enzyme regulating JH titer, downregulation via tomato expressing dsRNA disrupted the metamorphosis and adult emergence in *H. armigera* [75]. Similarly, silencing of chitinase mainly found in insect midgut, integument cell walls, cuticles, shells, and intestinal peritrophic matrices (PMs) play important role during insect molting and metamorphosis [76]. Continuous feeding of tomato transgenic leaves expressing hairpin RNA complimentary to chitinase gene of *H. armigera* led to reduced gene transcript which induced detrimental effect on the overall development and survival of insect [77]. Aphids (*Myzus persicae*) are sap-sucking pests that cause significant crop loss by direct feeding and transmitting the virus causing severe diseases in plants [78]. Tomato-mediated RNAi to silence acetylcholinesterases (AchE) which work as neurotransmitters in insects, resulted into reduced aphid fecundity [79]. The endogenous gene regulation pathway of miRNA is exploited by amiRNA technology to control the gene of interest and has shown significant silencing of the target gene with less or no off-target effects [80–83]. Silencing of ecdysone receptor gene (EcR), involved in all the stages of insect's life cycle, through tomato expressing amiRNA significantly increased the tolerance of plants towards insect attack [84].

Another popularly consumed vegetable, Potato (*Solanum tuberosum*) belongs to the family Solanaceae, and is the 4th most grown crop after wheat, rice and maize [85]. The crop is highly nutritious since it is rich in carbohydrates, proteins, minerals and vitamins [86]. Various biotic and abiotic stress factors limit the production and crop yield. Various biotic and abiotic stress factors limit the production and crop yield. Common potato infesting insects are Colorado potato beetle, potato tuber moth, green peach aphid (*M. persicae*), potato aphid, beet leaf hoppers, thirps and mites. Colorado potato beetle (CPB), *Leptinotarsa decemlineata*, is the most important pest due to the detoxification mechanism to survive various natural and synthetic chemicals [87]. RNAi- mediated silencing of *EcR* gene of CPB expressed in transgenic potato, showed 80% mortality and inability of the insect to complete the life cycle [88]. Similarly, feeding of transgenic potato expressing the hairpin RNA against JHAMT gene of CPB, led to reduced transcript level of the targeted gene and also significantly affected the growth and development of the pest specially the oviposition. Field trials of the transgenic potato showed high tolerance to the pest infestation and the surviving insects displayed low reproduction potential [89]. Potato transgenics encoding the RNAi construct targeting the host's gene Glycoalkaloid metabolism 4 (GAME4) coding for cytochrome P450, resulted into early instar mortality and accelerated insect development [90]. Similar to CPB, insect pest *Phthorimaea operculella* also causes huge losses to the production [91]. RNAi mediated control of insect pest has been demonstrated through topical application of dsRNA targeting Chitin Synthase A gene [92].

Other common vegetables like Cauliflower and Cabbage etc. are also heavily infested by insect pests. Cauliflower, belonging to species *Brassica oleracea* is profoundly infested by diamondback moth *Plutella xylostella*. It is one of the most destructive insect pests of Brassica all over the world for its short life span, high reproductive potential, lack of natural predators, and its ability to become resistant to a wide range of toxins and growth regulators [93, 94]. Cabbage, another member of the "cole" group crops is infested by many lepidopteran pests such as *P. xylostella, Pieris rapae, Mamestra brassicae* and *Trichoplusia ni* causing a major constraint of its yield [95, 96]. Various studies demonstrate the potential of RNAi mediated management of lepidopteran pest complex of cauliflower and cabbage, but dsRNA expressing transgenic plants targeting the pest complex have yet not been reported [97–100].

5. Conclusions

Post discovery, RNAi technology has been harnessed as a functional genomics tool as well as a crop improvement tool for various applications including control of insect pests. With its unique insecticidal mode of action, suppression of gene expression, it can act individually as well as can complement the current methods deployed for pest control. The technology has been applied in a range of crops against insect pests from orders such as Coleoptera, Lepidoptera, and Hemiptera. However, RNAi efficacy varies in insects for reasons like dsRNA molecule itself, instability of dsRNA due to presence of nucleases and gut pH, incomplete or impaired dsRNA internalization, lacking core RNAi machinery, weakened systemic spreading, developmental stage used for silencing and refractory target genes [100]. Therefore, for deployment of this technology on a commercial scale these challenges need to be addressed. In the course of evolution, insects are known for their remarkable adaption, allowing them to evolve resistance to any control method, including transgenic plants with protective traits like insecticidal proteins and RNAi. Thus, to provide sustainable crop protection managing the pest resistance issue, integrated pest management (IPM) approaches using combination of various control strategies like preventive measures like crop rotation, intercropping or cultivation of pest-resistant varieties, use of natural biocontrol factors such as pathogens or predators, and genetic control via transgenic plants expressing transgenes (Insecticidal proteins/dsRNA targeting insect genes) or release of sterile insects should be deployed.

Author details

Shipra Saxena[1], Sneha Yogindran[2], Manmohan Arya[3], Yogita Sharma[3] and Chandra Pal Singh[3]*

1 Department of Genetics, University of Delhi, New Delhi, India

2 Department of Biotechnology, Cochin University of Science and Technology, Kerala, India

3 Department of Botany, University of Rajasthan, Jaipur, Rajasthan, India

*Address all correspondence to: chandrapal203@gmail.com

References

[1] FAO. (2009). FAO's director-general on how to feed the world in 2050. Population Dev. Rev. 35, 837-839.

[2] Gautham H.R., Bhardwaj M.L. International Year of Plant Health 2020 to focus on threat of pests on food security. Curr. Sci. 2020;118:857.

[3] Deutsch CA, Tewksbury JJ, Tigchelaar M, Battisti DS, Merrill SC, Huey RB, et al.. Increase in crop losses to insect pests in a warming climate. Science (2018):919:916-919. doi: 10.1126/science.aat3466

[4] Mallet, J. Lepidoptera Taxome Project Draft Proposals and Information. Available online: http:// www.ucl.ac.uk/taxome/lepnos.html (accessed on 20 August 2016).

[5] Xu J, Wang X-F, Chen P, Liu FT, Zheng SC, Ye H, Mo MH (2016) RNA interference in moths: mechanisms, applications and progress. Genes 7:88. doi: 10.3390/genes7100088

[6] Zalucki, M.P.; Shabbir, A.; Silva, R.; Adamson, D.; Liu, S.S.; Furlong, M.J. Estimating the economic cost of one of the world's major insect pests, Plutella xylostella (Lepidoptera: Plutellidae): Just how long is a piece of string? J. Econ. Entomol. 2012, 105, 1115-1129. [Google Scholar] [CrossRef] [PubMed]

[7] Asaro, C. Chamberlin, L.A. Outbreak history (1953-2014) of spring defoliators impacting oak-dominated forests in Virginia, with emphasis on Gypsy Moth (*Lymantria dispar* L.) and Fall Cankerworm (*Alsophila pometaria* Harris). Am. Entomol. 2015, 61, 174-185. [Google Scholar] [CrossRef]

[8] Cox, P.D.; Bell, C.H. Biology and ecology of moth pests of stored foods. In Ecology and Management of Food-Industry Pests (FDA Technical Bulletin Number 4); Gorham, J.R., Ed.; The Association of Official Analytical Chemists: Gaithersburg, MD, USA, 1991; pp. 181-193. [Google Schol

[9] Choudhary B and Gaur K (2009) The Development and Regulation of Bt Brinjal in India (Eggplant/ Aubergine). ISAAA Brief No.38. ISAAA: Ithaca, NY

[10] Bravo, A., Gill, S. and Soberón, M. Mode of action of *Bacillus thuringiensis* Cry and Cyt toxins and their potential for insect control. Toxicon. 2007: 49: 423-435.

[11] Crickmore N, Zeigler DR, Schnepf E, Van Rie J, Lereclus D, et al. (2010) *Bacillus thuringiensis* toxin nomenclature. http://www. lifesci.sussex.ac.uk/ home/ Neil_ Crickmore/Bt/.

[12] Roh JY, Choi JY, Li MS, Jin BR, Je YH. Bacillus thuringiensis as a specific, safe, and effective tool for insect pest control. Journal Microbiology and Biotechnology. 2007; 17: 547-559.

[13] Crickmore N. Using worms to better understand how *Bacillus thuringiensis* kills insects. Trends in Microbiology 2005; 13: 347-350.

[14] van Frankenhuyzen K. Cross-order and cross-phylum activity of *Bacillus thuringiensis* pesticidal proteins. Journal of Invertebrate Pathology. 2013; 114: 76-85.

[15] Ranjekar PK, Patankar A, Gupta V, Bhatnagar R, Bentur J, Kumar PA Genetic engineering of crop plants for insect resistance. Current Science. 2003; 84: 321-329.

[16] Sheikh AA, Wani MA, Bano P, Sajad Un Nabi SU, et al. An overview on resistance of insect pests against Bt Crops. Journal of Entomology and Zoology Studies. 2017; 5: 941-948.

[17] Allah B, Emine A, Delpasand KS, Cem KO, Cengiz S, Sebahattin O. Development of insect-resistant cotton lines with targeted expression of insecticidal gene. Archives of Biological Sciences. 2016; 68: 773-780.

[18] Bates SL, Zhao J-Z, Roush RT, Shelton AM. Insect resistance management in GM crops: past, present and future. Nature Biotechnology. 2005; 23: 57-62.

[19] Srivastava DK, Kumar P, Sharma S, Gaur A, and Gambhir G (2016) Genetic Engineering for Insect Resistance in Economically Important Vegetable Crops M. Anis, N. Ahmad (eds.), Plant Tissue Culture: Propagation, Conservation and Crop Improvement, DOI 10.1007/978-981-10-1917-3_15

[20] Stevens J, Dunse K, Fox J, Evans S and Anderson M (2012) Biotechnological approaches for the control of insect pests in crop plants. In: Pesticides-advances in chemical and botanical pesticides, (Soundararajan RP, ed.), ISBN. 978-953-51-0680-7, In: Tech, doi: 10.5772/46233.

[21] Tabashnik BE, Carrière Y. Surge in insect resistance to transgenic crops and prospects for sustainability. Nature Biotechnology. 2017 Oct;35(10):926.

[22] Saurabh S, Vidyarthi AS, Prasad D. RNA interference: concept to reality in crop improvement. Planta. 2014 Mar;239(3):543-64. Katoch R, Sethi A, Thakur N, et al. RNAi for insect control: Current perspective and future challenges. Applied Biochemistry and Biotechnology. 2013; 171: 847-873.

[23] Yogindran S, Rajam MV (2015) RNA interference strategy for crop protection against insect pests. In: Bt resistance, characterization and strategies for GM crops producing *Bacillus thuringiensis*. Toxins. Soberon M, Gao Y, Bravo A (eds.) CABI Biotechnology Series 4. CAB International, Oxfordshire, pp. 162-172.

[24] Mello CC, Conte D Jr. Revealing the world of RNA interference. Nature. 2004 Sep 16;431(7006):338-42. doi: 10.1038/nature02872. doi: 10.1038/nature02872.

[25] Fire, A., Xu, S., Montgomery, M. K., Kostas, S. A., Driver, S. E., and Mello, C. C. Potent and specific genetic interference by double-stranded RNA in *Caenorhabditis elegans*. Nature. 1998; 391, 806-811. doi: 10.1038/35888.

[26] Baulcombe, D.C. RNA silencing in plants. Nature. 2004; 431, 356-363. doi: 10.1038/nature02874.

[27] Wilson RC, Doudna JA. Molecular mechanisms of RNA interference. Annu Rev Biophys. 2013;42:217-39. doi: 10.1146/annurev-biophys-083012-130404. doi: 10.1146/annurev-biophys-083012-130404.

[28] Meister G, Tuschl T. Mechanisms of gene silencing by double-stranded RNA. Nature. 2004 Sep 16;431(7006):343-9. doi: 10.1038/nature02873.

[29] Bernstein E, Caudy AA, Hammond SM, Hannon GJ (2001). "Role for a bidentate ribonuclease in the initiation step of RNA interference". Nature. 409 (6818): 363-6. doi: 10.1038/35053110.

[30] Margis R, Fusaro AF, Smith NA, Curtin SJ, Watson JM, Finnegan EJ, Waterhouse PM. The evolution and diversification of Dicers in plants. FEBS Lett. 2006 May 1;580(10):2442-50. doi: 10.1016/j.febslet.2006.03.072. doi: 10.1016/j.febslet.2006.03.072.

[31] Agrawal N, Dasaradhi PV, Mohmmed A, Malhotra P, Bhatnagar RK, Mukherjee SK. RNA interference: biology, mechanism, and applications. Microbiol Mol Biol Rev.

2003 Dec;67(4):657-85. doi: 10.1128/mmbr.67.4.657-685.2003.

[32] Baum JA, Roberts JK. Progress towards RNAi-mediated insect pest management. Adv Insect Physiol. 2014;47:249-295. doi: 10.1016/B978-0-12-800197-4.00005-1.

[33] Burand JP, Hunter WB. RNAi: future in insect management. J Invertebr Pathol. 2013;112:S68–S74. doi: 10.1016/j.jip.2012.07.012.

[34] Mamta B, Rajam MV. RNAi technology: a new platform for crop pest control. Physiol Mol Biol Plants. 2017;23(3):487-501. doi:10.1007/s12298-017-0443-x.

[35] Liu S, Jaouannet M, Dempsey DA, Imani J, Coustau C, Kogel KH. RNA-based technologies for insect control in plant production. Biotechnol Adv. 2020 Mar-Apr;39:107463. doi: 10.1016/j.biotechadv.2019.107463.

[36] Scott JG, Michel K, Bartholomay LC, Siegfried BD, Hunter WB, Smagghe G, Zhu KY, Douglas AE (2013) Towards the elements of successful insect RNAi. J Insect Physiol 59:1212-1221. doi: 10.1016/j.jinsphys.2013.08.014

[37] Baum JA, Bogaert T, Clinton W, Heck GR, Feldmann P, Ilagan O, Johnson S, Plaetinck G, Munyikwa T, Pleau M, Vaughn T. Control of coleopteran insect pests through RNA interference. Nature biotechnology. 2007 Nov;25(11):1322-6.

[38] Hunter WB, Glick E, Paldi N and Bextine BR. Advances in RNA interference: dsRNA treatment in trees and grapevines for insect pest suppression. South West Entomology 2012; 37(1):85-87.

[39] Dalakouras A, Jarausch W, Buchholz G, Bassler A, Braun M, Manthey T, Krczal G, Wassenegger M. Delivery of hairpin RNAs and small RNAs into woody and herbaceous plants by trunk injection and petiole absorption. Frontiers in Plant Science. 2018; 24;9:1253.

[40] Ghosh SKB, Hunter WB, Park AL. Gundersen-Rindal Double strand RNA delivery system for plant-sap-feeding insects. PLoS One. 2017; 12: e0171861

[41] Joga MR, Zotti MJ, Smagghe G and Christiaens O. RNAi efficiency, systemic properties, and novel delivery methods for pest insect control: what we know so far. Front Physiol 2016; 7:553.

[42] Chen X, Li L, Hu Q, Zhang B, Wu W, Jin F et al. Expression of dsRNA in recombinant Isaria fumosorosea strain targets the TLR7 gene in *Bemisia tabaci*. BMC Biotechnol. 2015; 15: 64.

[43] Kumar P, Pandit SS, Baldwin IT. Tobacco rattle virus vector: a rapid and transient means of silencing manduca sexta genes by plant mediated RNA interference. PLoS One, 2012; 7: e e31347

[44] Tian, H., Peng, H., Yao, Q., Chen, H., Xie, Q., Tang, B. and Zhang, W. Developmental control of a Lepidopteran pest *Spodoptera exigua* by ingestion of bacterial expressing dsRNA of a non-midgut gene. PLoS ONE, 2009; 4: e6225

[45] Kolliopoulou A, Taning CNT, Smagghe G, Swevers L. Viral Delivery of dsRNA for Control of Insect Agricultural Pests and Vectors of Human Disease: prospects and Challenges. Front. Physiol. 2017; 8: 1-24

[46] Cagliari D, Dias NP, GALDEANO D, dos Santos EÁ, Smagghe G and Zotti MJ. Management of pest insects and plant diseases by non-transformative RNAi. Frontiers in plant science. 2019; 10:1319.

[47] Li H, Khajuria C, Rangasamy M, Gandra P, Fitter M, Geng C, Woosely A, Hasler J, Schulenberg G, Worden S, McEwan R. Long ds RNA but not si RNA initiates RNA i in western corn

rootworm larvae and adults. Journal of Applied Entomology. 2015 Jul;139(6):432-45.

[48] Zhang J, Khan SA, Hasse C, Ruf S, Heckel DG & Bock R. Full crop protection from an insect pest by expression of long double-stranded RNAs in plastids. Science 2015; 347: 991-994.

[49] de la Fuente J, Kocan KM, Almazán C and Blouin EF. RNA interference for the study and genetic manipulation of ticks. Trends in Parasitology. 2007; 23(9):427-433.

[50] Duan CG, Wang CH, Guo HS. Application of RNA silencing to plant disease resistance. Silence. 2012; 3(1): 5.

[51] Shaner DL and Beckie HJ. The future for weed control and technology. Pest Management Science 2014; 70(9): 1329-1339.

[52] Wang M, Weiberg A, Lin F-M, Thomma BPHJ, Huang H-D, Jin H. Bidirectional crosskingdom RNAi and fungal uptake of external RNAs confer plant protection. Nature Plants. 2016; 2(10):16151

[53] San Miguel K and Scott JG. The next generation of insecticides: DsRNA is stable as a foliar-applied insecticide. Pest Management Science. 2016; 72(4): 801-809.

[54] Zotti M, dos Santos EA, Cagliari D, Christiaens O, Taning CNT and Smagghe G. RNA interference technology in crop protection against arthropod pests, pathogens and nematodes. Pest Management Science. 2018; 74(6): 1239-1250.

[55] Niu J, Shen G, Christiaens O, Smagghe G, He L, Wang J. Beyond insects: current status and achievements of RNA interference in mite pests and future perspectives. Pest management science. 2018 Dec;74(12):2680-7.

[56] Head GP, Carroll MW, Evans SP, Rule DM, Willse AR, Clark TL, Storer NP, Flannagan RD, Samuel LW, Meinke LJ. Evaluation of SmartStax and SmartStax PRO maize against western corn rootworm and northern corn rootworm: efficacy and resistance management. Pest management science. 2017; 73(9):1883-99.

[57] Waltz, E. (2015). USDA approves next-generation GM potato. Nat. Biotechnol. 33, 12-13.

[58] Baranski, R., Klimek-Chodacka, M., Lukasiewicz, A. (2019). Approved genetically modified (GM) horticultural plants: a 25-year perspective. Folia Hortic. 31, 3-49.

[59] Mao YB, Cai WJ, Wang JW, Hong GJ, Tao XY, Wang LJ, Huang YP and Chen XY. Silencing a cotton bollworm P450 monooxygenase gene by plant-mediated RNAi impairs larval tolerance of gossypol. Nature Biotechnology. 2007; 25 (11):1307-1313.

[60] Dhadialla TS, Carlson GR, Le DP. New insecticides with ecdysteroidal and juvenile hormone activity. Annu Rev Entomol. 1998;43:545-569

[61] Zhu JQ, Liu S, Ma Y, Zhang JQ, Qi HS, Wei ZJ, Yao Q, Zhang WQ, Li S (2012) Improvement of pest resistance in transgenic tobacco plants expressing dsRNA of an insect associated gene EcR. PLoS ONE 7:e38572

[62] Xiong Y, Zeng H, Zhang Y, Xu D, Qiu D. Silencing the HaHR3 gene by transgenic plant-mediated RNAi to disrupt Helicoverpa armigera development. Int J Biol Sci 2013:9:370-381

[63] Liu F, Wang XD, Zhao YY, Li YJ, Liu YC, Sun J (2015) Silencing the HaAK gene by transgenic plant-mediated RNAi impairs larval growth of Helicoverpa armigera. Int J Biol Sci 11:67-74

[64] Jin S, Singh ND, Li L, Zhang X, Daniell H (2015) Engineered chloroplast dsRNA silences cytochrome p450 monooxygenase, V-ATPase and chitin synthase genes in the insect gut and disrupts Helicoverpa armigera larval development and pupation. Plant Biotech J 13:435-446

[65] Zha WJ, Peng XX, Chen RZ, Du B, Zhu LL, He GC (2011) Knockdown of midgut genes by dsRNA-transgenic plant-medi- ated RNA interference in the Hemipteran insect *Nilaparvata lugens*. Plos One 6:e20504

[66] Poreddy, S., Li, J. & Baldwin, I.T. Plant-mediated RNAi silences midgut-expressed genes in congeneric lepidopteran insects in nature. BMC Plant Biol 2017:17:199. doi.org/10.1186/s12870-017-1149-5

[67] Jiang S, Wu H, Liu H, Zheng J, Lin Y, Chen H. The overexpression of insect endogenous small RNAs in transgenic rice inhibits growth and delays pupation of striped stem borer (Chilo suppressalis). Pest management science 2017:73(7):1453-1461.

[68] Zheng, X., Weng, Z., Li, H., Kong, Z., Zhou, Z., Li, F., ... & Chen, H. (2020). Transgenic rice overexpressing insect endogenous microRNA csu-novel-260 is resistant to striped stem borer under field conditions. Plant Biotechnology Journal.

[69] Baum JA, Bogaert T, Clinton W, Heck GR, Feldmann P, Ilagan O, Johnson S, Plaetinck G, Munyikwa T, Pleau M, Vaughn T. Control of coleopteran insect pests through RNA interference. Nature biotechnology. 2007 Nov;25(11):1322-6.

[70] Meng, F., Li, Y., Zang, Z., Li, N., Ran, R., Cao, Y., ... & Li, W. (2017). Expression of the double-stranded RNA of the soybean pod borer Leguminivora glycinivorella (Lepidoptera: Tortricidae) ribosomal protein P0 gene enhances the resistance of transgenic soybean plants. Pest management science, 73(12), 2447-2455.

[71] Wang, Z., Li, T., Ni, H., Wang, G., Liu, X., Cao, Y., ... & Meng, F. (2018). Transgenic soybean plants expressing Spb18S dsRNA exhibit enhanced resistance to the soybean pod borer Leguminivora glycinivorella (Lepidoptera: Olethreutidae). Archives of insect biochemistry and physiology, 98(2), e21461.

[72] Mao, YB., Tao, XY., Xue, XY. et al. Cotton plants expressing CYP6AE14 double-stranded RNA show enhanced resistance to bollworms. Transgenic Res 20, 665-673 (2011).

[73] Selvanarayanan V, Narayanasamy P Factors of resistance in tomato accessions against the fruit worm, Helicoverpa armigera (Hubner). Crop Prot 2006; 25:1075-1079

[74] Maligeppagol M, Asokan R, Krishna V, Latha J, NK KK, Ellango R. Transgenic tomato expressing dsRNA of juvenile hormone acid O-methyl transferase gene of *Helicoverpa armigera* (Lepidoptera: Noctuidae) affects larval growth and its development. Journal of Asia-Pacific Entomology. 2017 Jun 1;20(2):559-67.

[75] Agrawal N, Sachdev B, Rodrigues J, Sree KS, Bhatnagar RK. Development associated profiling of chitinase and microRNA of Helicoverpa armigera identified chitinase repressive microRNA. Scientific reports. 2013 Jul 26;3:2292.

[76] Mamta, K. R. K. Reddy, M. V. Rajam. Targeting chitinase gene of Helicoverpa armigera by host-induced RNA interference confers insect resistance in tobacco and tomato. Plant Mol Biol 2016; 90:281-292

[77] Hogenhout, S.A.; Ammar, E.-D.; Whitfield, A.E.; Redinbaugh, M.G. Insect Vector Interactions with

Persistently Transmitted Viruses. Annu. Rev. Phytopathol. 2008; 46: 327-359

[78] Faisal M, Abdel-Salam EM, Alatar AA, Saquib Q, Alwathnani HA, Canto T. Genetic Transformation and siRNA-Mediated Gene Silencing for Aphid Resistance in Tomato. Agronomy. 2019 Dec;9(12):893.

[79] Pareek M, Yogindran S, Mukherjee SK, Rajam MV. Plant microRNAs: Biogenesis, functions, and applications. InPlant biology and biotechnology 2015 (pp. 639-661). Springer, New Delhi.

[80] Niu QW, Lin SS, Reyes JL, Chen KC, et al. Expression of artificial microRNAs in transgenic *Arabidopsis thaliana* confers virus resistance. Nature Biotechnology 2006; 24: 1420-1428.

[81] Schwab R, Ossowski S, Riester M, Warthmann N, Weigel D. Highly specific gene silencing by artificial microRNAs in Arabidopsis. Plant Cell 2006; 18: 1121-1133

[82] Alvarez JP, Pekker I, Goldshmidt A, Blum E, Amsellem Z, Eshed Y. Endogenous and synthetic microRNAs stimulate simultaneous, efficient, and localized regulation of multiple targets in diverse species. Plant Cell. 2006: 18: 1134-1151.

[83] Yogindran S and Rajam MV. Host-derived artificial miRNA-mediated silencing of ecdysone receptor gene provides enhanced resistance to *Helicoverpa armigera* in tomato. Genomics. 2020 Oct 12.

[84] FAOSTAT data (2017) http://www.fao.org/home/en/. Retrieved 30 July 2018

[85] Calıskan ME, Onaran H, Arıoğ lu H. Overview of the Turkish potato sector: challenges, achievements and expectations. Potato Res 2010; 53:255-266

[86] Zhu F, Xu JJ, Palli R, Ferguson J, Palli SR. Ingested RNA: interference for managing the populations of the Colorado potato beetle (*Leptinotarsa decemlineata*). Pest Manag Sci. 2011; 67:175-182

[87] Hussain T, Aksoy E, Calıskan ME, Bakhsh A. Transgenic potato lines expressing hairpin RNAi construct of molting-associated EcR gene exhibit enhanced resistance against Colorado potato beetle (*Leptinotarsa decemlineata*, Say). Transgenic Res. 2019; 28:151-164

[88] Guo W, Bai C, Wang Z, Wang P, Fan Q, Mi X, Wang L, He J, Pang J, Luo X, Fu W. Double-stranded RNAs high-efficiently protect transgenic potato from *Leptinotarsa decemlineata* by disrupting juvenile hormone biosynthesis. Journal of agricultural and food chemistry. 2018 Nov 6;66(45):11990-9.

[89] Paudel JR, Davidson C, Song J, Maxim I, Aharoni A, Tai HH. Pathogen and Pest Responses Are Altered Due to RNAi-Mediated Knockdown of GLYCOALKALOID METABOLISM 4 in *Solanum tuberosum*, MPMI. 2017; 30(11): 876-885.

[90] Rondon, SI. The potato tuberworm: a literature review of its biology, ecology, and control. American Journal of Potato Research 2010; 87: 149-166.

[91] Mohammed AM, Diab MR, Abdelsattar M, Sayed MS. Characterization and RNAi-mediated knockdown of Chitin Synthase A in the potato tuber moth, *Phthorimaea operculella*. Scientific Reports. 2017 Aug 25;7(1):1-2.

[92] Furlong MJ, Wright DJ, Dosdall LM. Diamondback moth ecology and management: problems, progress, and prospects. Annual review of entomology. 2013 Jan 7;58:517-41.

[93] Talekar NS, Shelton AM. Biology, ecology, and management of the diamondback moth. Annual review of entomology. 1993 Jan;38(1):275-301.

[94] Iqbal, M., Khan, A.R., Rafique, M., Suthar, V. and Solangi, B.K. Comparative Efficacy of Botanical and Synthetic Pesticides against Major Insect Pests of Cabbage. European Academic Research, 2015; 3(9):10038-10057.

[95] Cartea, M. E.; Francisco, M.; Lema, M.; Soengas, P.; Velasco, P. Resistance of Cabbage (*Brassica oleracea* capitata Group) Crops to Mamestra brassicae. Journal of Economic Entomology, 2010; 103(5), 1866-1874.

[96] Israni B and Rajam MV. Silencing of ecdysone receptor, insect intestinal mucin and sericotropin genes by bacterially produced double-stranded RNA affects larval growth and development in Plutella xylostella and Helicoverpa armigera. Insect Mol Biol, 2017; 26: 164-180.

[97] Wang Y, Liu Z, Xu J, Li X, Bi H, Andongma AA, Niu C, Huang Y. Mutation of doublesex induces sex-specific sterility of the diamondback moth Plutella xylostella. Insect biochemistry and molecular biology. 2019 Sep 1;112:103180.

[98] Ellango R, Asokan R, Chandra GS, Kumar NK, Mahmood R, Ramamurthy VV. Tyrosine hydroxylase, a potential target for the RNAi-mediated management of diamondback moth (Lepidoptera: Plutellidae). Florida Entomologist. 2018 Mar;101(1):1-5.

[99] Su Liu, Yu-Xing Zhang, Wen-Long Wang, Bang-Xian Zhang, Shi-Guang Li. Identification and characterisation of seventeen glutathione S-transferase genes from the cabbage white butterfly *Pieris rapae*, Pesticide Biochemistry and Physiology, 2017; 43:102-110.

[100] Cooper AM, Silver K, Zhang J, Park Y, Zhu KY. Molecular mechanisms influencing efficiency of RNA interference in insects. Pest management science. 2019 Jan;75(1):18-28.

Chapter 3

Role of Pheromone Application Technology for the Management of Codling Moth in High Altitude and Cold Arid Region of Ladakh

Barkat Hussain, Faizaan Ahmad, Ejaz Ahmad, Wasim Yousuf and Mohd Mehdi

Abstract

The codling moth is a threat to the apple industries in India. Currently, no solutions are available for the management of codling moth in Ladakh. Therefore, all fresh fruits from Ladakh are still banned due to quarantine regulations. Jammu and Kashmir and Himachal Pradesh and Ladakh are the three main apple producing states of India, both in quality and quantity. The ban on all fresh fruits from Ladakh directly affects the economy of rural populations. These fruits are sold in all the local markets of Kargil and Leh. Apples damaged by the larvae of codling moth are less preferred by inhabitants, tourists, and security forces, a large area of Ladakh is bordered with China and Pakistan. Field demonstration trials revealed significantly less fruit damage in apple orchards in different hamlets of Ladakh using pheromone dispensers, pheromone baited traps, and two applications of insecticides for codling moth management. A demonstration of the use of pheromone and pheromone dispenser technology for area-wide management for high dense populations of the codling moth in Ladakh has revealed successful results in the orchards of the apple growers. Area-wide management of the codling moth in some villages, using dispenser technology has shown promising results. The ban of fresh fruits in Ladakh may not be, therefore, appropriate as management of the codling moth appears to be successful with the use of pheromone dispenser technology. This technology will, surely, boost the apple industry and have a great potential for establishing commercial orchards and quality apples in high altitudes in the second-highest cold arid region of the world.

Keywords: pheromone technology, codling moth, Ladakh, pheromone dispensers, cold arid region

1. Introduction

In India, apple (*Malus domestica* Borkh) are the main dominating fruit crop among all the cultivated fruits grown in Jammu and Kashmir. Annually production is 17.268 lakh metric tons of apple fruits cultivated in an area of 1.629 lakh hectares. The other apple-producing states of India are Himachal Pradesh, Uttarakhand, and some parts of Arunachal Pradesh [1]. Jammu and Kashmir (J&K) account for 60%

of the total apple production among all the cultivated fruits before splitting the state into two Union territories, namely J&K and Ladakh [2, 3]. Ladakh is the second highest, cold arid region of the world. The low precipitation and scarce vegetation over a larger area make Ladakh a high-altitude desert. Ladakh is on the eastern side is connected with China. Pakistan is on the northwest border and to the west is the Kashmir Valley and Himachal Pradesh is on the southern border. Ladakh was extended by the Great Himalaya to the south and the Karakoram mountains to its north. Ladakh is administratively divided into two districts, Leh District and Kargil District. Ladakh was a separate province of Jammu and Kashmir State, constituting the major part of the State until 2019. Recently, in August 2019, the parliament of India passed an act and Ladakh is now being administered as a Union territory of India [4]. Ladakh was considered an important trade link in the past, but border closure with China restricts the movement of all essential commodities and other fresh fruits. The fruit is now being airlifted during the winter which increases the prices of these commodities.

The new Union territory of Ladakh produces annually 3241 metric tonnes of apples from an area of 598 hectares [5]. The dominating fruit crop of Ladakh is apricot grown over an area of 2127 hectares with an annual production of 12686 metric tonnes (**Table 1**). Other important fruit crops (pear, plum, cherry, grapes, and walnut) are grown in areas of 243 ha with an annual production of 319 metric tonnes. The area and production of these fruit crops are detailed in the pie chart (**Figures 1** and **2**). After calculating the area and production of Ladakh under fruit crops, the total fruit crops are grown in an area of 2968 ha with an annual production of 16246 metric tonnes, as reported by the National Horticulture Board of India. In India, the codling moth is found only in Ladakh [6, 7] and, other parts of India are free from this pest. Therefore, the attack of this single pest restricts the movement or transportation of all fresh fruit from Ladakh to other states of India. The monetary losses incurred on the ban on all fresh fruits from Ladakh could be roughly estimated as more than 10,000 Cr INR annually after calculating the prices of these fruit crops in national and international markets by multiplying the huge production of 16246 metric tonnes of fruit crops produced annually [8, 9].

The extent of fruit damage on apples and apricots in Ladakh, caused by the codling moth, is so high that it can enter easily into any neighboring states. Therefore, the Government of India prohibited the export of all fresh fruits from Ladakh within or outside the state under the Law" Destructive Insect Pest Act 1914 (II) before splitting the Jammu and Kashmir state into two union territories. This ban has been strictly imposed by the enforcement agencies to prevent its entry into the neighboring states of Jammu and Kashmir, Himachal Pradesh, Uttar Pradesh, Uttrakhand, and other apple growing areas of India. Such ban on all fresh fruits have prevented also its entry to other parts of the Indian union till to this date. Therefore the ban on the export of all fresh fruits from Ladakh is a good rationale.

Ladakh	Area (ha)	Production (MT)
Apple	598	3241
Apricot	2127	12686
Other Fruits	243	319
Total Production	2968	16246

Table 1.
Total area and production of main fruit crops in Ladakh during 2018–2019.

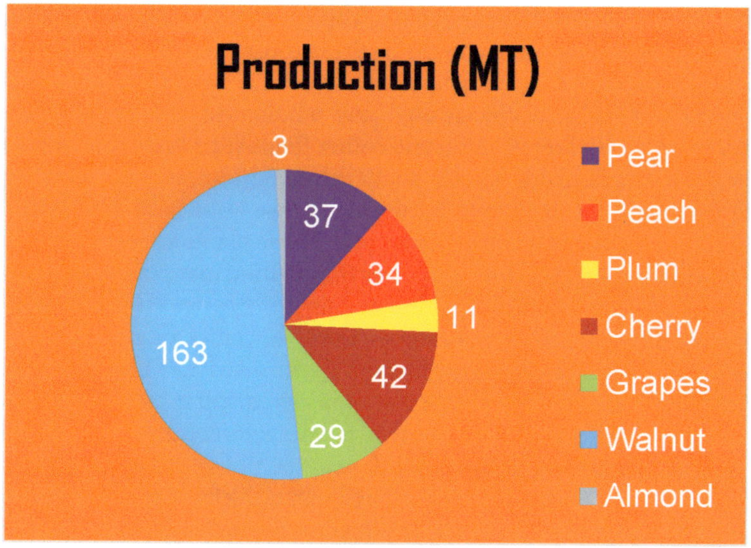

Figure 1.
Detailed production of other fruit crops in Ladakh 2018–2019.

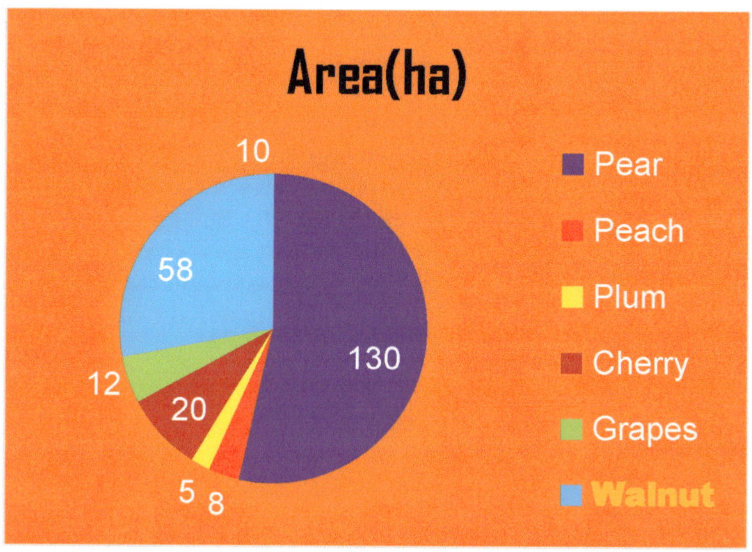

Figure 2.
Detailed area of other fruit crops in Ladakh 2018–2019.

The codling moth, *Cydia pomonella* L, is a well-known notorious pest globally and has been reported from Europe, USA, Canada, South Africa, Australia, New Zealand, Afghanistan, Iran, South Africa, Pakistan, Germany, France, Russia, and other apple-growing regions of the globe [10]. In India, its distribution is restricted to Ladakh [11, 12]. The codling moth is thought to have entered Ladakh through the Northwest Frontier province of Pakistan [12–14]. The infestation level of the codling moth on apple, apricot, pear, and walnut has been observed in all the fruit-growing areas of Ladakh [14]. While, [15] reported that the fruit damage, caused by codling moth, ranged from 42.5 to 49.7 percent in Leh and Kargil, respectively. The codling moth is a single pest on apple in Ladakh and is found to attack all the local and introduced cultivars of apple [14]. Recent surveys conducted in different hamlets of Ladakh revealed that the level of fruit infestation ranged from 70.0–83.0% (**Table 2**). The extent of fruit damage on other fruit crops caused by the codling moth has not been estimated. Mostly, in all the villages of Ladakh, apple plants are been raised in isolated patches of land. Due to the small size of cultivated land holdings in fruit-producing areas of Ladakh, apple trees are planted along with other fields such as cereals, vegetables, or field crops. The latter field crops act as shelterbelts. Besides, to the cold arid climate and scarcity of water, raising plants in this region is very difficult. Moreover, the apple trees in Ladakh are of the spreading type. These plants are kept untrained, un-pruned, and un-managed due to scarcity of qualified manpower, less awareness, social taboos, and other poor information among the growers [16]. The region falls under Tribal Region. In such circumstances, the insecticides are not effective in the Kargil district, heavily populated by Muslims. In the district of Leh, the majority of the population are Buddhists. In Buddhism, it is a sin to kill insects or any type of pests (boring/sucking/feeding) due to socio-religious constraints. In such situations, the pest populations and the level of fruit damage of all insect pests are increasing at an alarming rate and are being considered a threat to the fruit industry [5, 16].

The codling moth adults are grayish-brown in color, one inch long, and bear coppery wings that have copper colouration at the tip region. The adult females lay eggs on fruits and leaf surfaces. From the hatched eggs, newly emerged larvae are white with blackheads, while the late instars larvae are light pink in color with blackheads. Codling moth larvae enter the apple fruit at the calyx end and directly bore into the fruit (**Figure 3(A)**) Later on, these larvae feed on fruit pulp and seeds. The fully-grown larvae move out from the apple fruits for pupation. During winter, larvae overwinter under loose or dead bark of apple trees (**Figure 3(B)**). Larvae overwinter in winter, pupates in spring, and emerge as adults from mid-May to mid-June in Ladakh [17]. Adult (egg to adult) emergence synchronized often with the peanut stage of the apple fruit as both the phenological stages of plant and insects require certain amounts of heat requirements [18–20]. Being a direct pest, larvae bore directly inside the fruits, but the early fruit damage is responsible for fruit fall and fruit injury. Larvae inside the fruit feed on the seeds and create exit holes and waste material is pushed out. Sometimes, the waste material remains associated with or inside the fruit that causes fruit rotting.

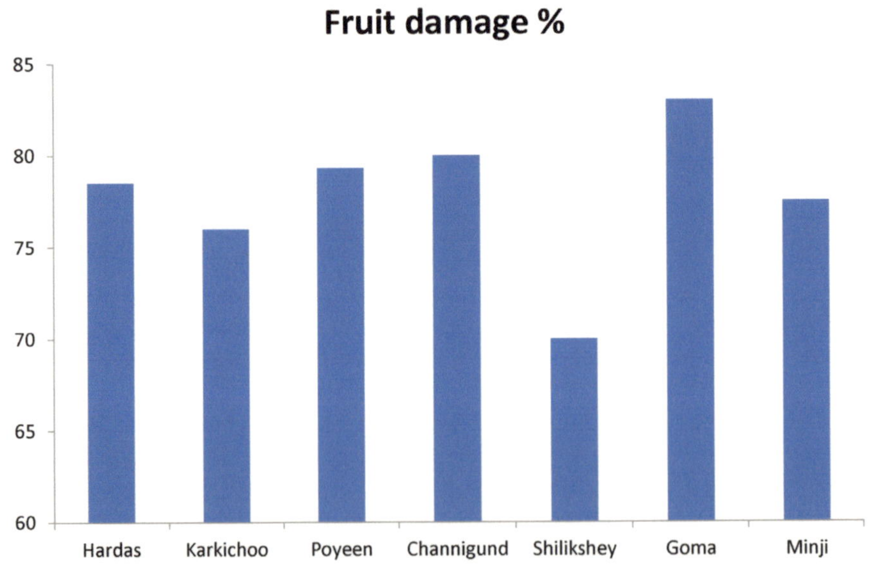

Table 2.
The level of fruit damage (%) in surveyed orchards of Kargil during 2018–2019.

Figure 3.
(A) Overwintering larvae exposed on loose bark, (B) bored fruits by codling moth, (C) pheromone dispenser inserted on apple branch, (D) trap catch of codling moth.

The boring habit of the larvae inside the fruits is not fit for consumption and is disliked by consumers and discarded for marketing purposes. Therefore, the economic thresh hold level (ETL) for this pest is very low. The codling moth is a quarantine pest and is of great importance concerning its quarantine requirement for trade purposes. To export apple fruits to any part of the world, stringent phytosanitary and other quarantine measures are warranted for exporting from reporting countries (presence of codling moth) to non-reporting (absence of codling Moth) countries. Sometimes trade is declined due to incursion of this pest where it has not been reported.

2. Pheromones for the management of codling moth

Insecticides are considered as entomological weapons against a wide group of insect pests. The insecticides are cheaper, wide spectrum, hazardous to consumers, and responsible for the quick kill to both codling moth and its natural enemies [21, 22]. The application of insecticides on fruit orchards lead to poisoning of the environment, eliminating the non-target organisms, insecticidal resistance, resurgence and secondary pest outbreaks [21, 22]. Indiscriminate application of insecticides, like organophosphates and carbamates against codling moth, has to lead to insecticidal resistance, death of natural enemies [21, 23]. Insects release specific chemical odors to attract their partners known as sex pheromones. Besides, pheromones are naturally produced, environmentally acceptable, species-specific chemical compounds and are considered a good choice for pest management [24, 25]

Pheromones are considered as the essential components for monitoring, mass trapping, mating disruption, attract and kill and also combined with other non-chemical strategies for the management of codling moth using pheromone technology. Pheromone disruption technology employs different application methods (Puffers, metered backpack sprayers, caulking guns, dosing guns, aircraft, rings, and SPLAT) to disrupt mating of codling moth [26–29] in various countries. Pheromone for the codling moth was identified as (E, E)-8,10-dodecadien-1-ol [30]. The other minor compounds as (E, Z)-8,10-dodecadien-1-ol, and 1-tetradecanol) increase the efficacy and also the behavior of the codling moth [31]. These chemicals have now been synthesized and used for lure making and monitoring the adult population of the codling moth. Pheromone dispensers have been widely used for the management of codling moth across the world [32, 33]. The estimated area in hectares (77,000) for North America, 38,000 for European Union, 19,000 for South Africa and 28,000 for Argentina, Chile, Australia, New Zealand, and Israel), employed for mating disruption for codling moth [27, 34]

Pheromone-driven technology is an important method for the management of the codling moth [27, 35]. The learning of sex communication in insects, mostly through chemicals has been utilized for the management of the codling moth [27] and other insects [36, 37]. Using the mating disruption technique, species-specific chemicals are released in greater quantities in the environment to disrupt mating [38–40] of the target pest. Pheromone dispenser technology has been demonstrated in more than 7000 hectares in the world [27], and is more powerful than male annihilation technique and mass trapping [17, 38]. Male annihilation technique (MAT) is an insect control method to reduce the male population using sext attractants or sex pheromones from a large area to disrupt sexual communication. Mass trapping is an acceptable method, when both male and female partners are trapped from a large area using both sex pheromones and plant volatiles and mostly adopted for eradication programmes. To our knowledge, SKUAST-K is the only university to fabricate and prepare the lures for codling moth and other insect pests in India. In addition, pheromone dispensers has not been registered against any insect pests in India to cut down the pesticide usage, safe environment, reduce health hazards, pollution in the environment and shall be helpful for the management of various insect pests on apple or other crops. The various management options already available in making integrated pest management programme safer and sustainable for our future generations has been fully documented through the use of mating dispenser technology for the codling moth in all developed countries [27, 39].

3. What are the gaps in the management of codling moth in Ladakh?

3.1 Pruning and training

Pruning and training are not practiced in Ladakh and emphasis should be given to pruning and training in apple orchards. Such practices do not allow the insecticidal spray to provide the best coverage on the fruit and the whole plant canopy. Moths, larvae, and eggs are less likely to come into contact with poisoned fruit surfaces. The unpruned plants are not managed properly to perform various orchard management operations. Such practices are responsible for bushy and alternate fruit bearing plants, and poor harvest. Due to no fruit thinning, the fruits that developed are small in size and poor in quality and quantity. Being, an arid climate, no fungal diseases have been reported on apple orchards as fungal diseases perpetuate and perform well in humid climatic conditions. In both the districts of Ladakh, the plants are kept untrained and the heights of these plants are so high and are very difficult for the power-driven motor sprayers to cover the whole plant. The scarcity of skilled pruners is being felt in Ladakh and efforts should be made to introduce the course certification for training and pruning, leading to employment benefits to local populations.

4. Orchard types

Very few commercial apple orchards were observed during our surveys in Ladakh. Besides, more than 90% of apple trees are grown as backyard orchards, and very difficult to use the insecticidal application. Under the canopy of backyard orchards of apple, different types of vegetables and fruit crops are being cultivated. It is very difficult and tricky to use insecticides under such circumstances. The reason for raising mixed farming is due to the scarcity and non-availability of water in cold arid climates. The plant to plant and row to row distance between apple plants is improper in Ladakh. The apple orchards are scattered and are not connected with roads. It is very difficult for the growers to perform various intercultural operations. Mostly apple orchards in Ladakh, established as backyard orchards. The other fruit crops (pear, walnut, and apricot) are preferred hosts for codling moth. The future strategy for the management of codling moth in these backyard orchards should be directed to all fruit crops which shelter codling moth population. The demarcation line or walls between the houses are erected of stones provide shelter for the colding moth immatures (summer and overwintered larvae). So the overwintered larvae of the codling moth get easy access for overwintering inside the crevices and gaps among the stone walls. The levels of infestation of the codling moth remain high. Besides, the awareness campaigns should be launched for the management of codling moth, orchard management and quality plants with deep rooted, root stocks of apple.

4.1 Orchard sanitation

Due to the high population density of the codling moth, infested fruits drop prematurely from the apple trees before maturity. Picking and removing these infested fruits is not carried out to break the cycle of codling moth infestation. It has been observed that due to the huge infestation of the codling moth, fruits are not being harvested and the larvae mostly overwinter in such apple fruits and also in the fallen fruits. The targeted burlapping for trapping the overwintering larvae without impregnating the wrapping material with insecticides is not effective. It is therefore recommended to use the insecticide impregnated materials for trapping and killing the overwintered larvae. It was observed during the surveys that the larvae of the codling moth still like to overwinter on the loose bark, stone bunds raised around the houses rather than on the burlapping materials wrapped around the stems and limbs of the old trees. Therefore, it is recommended to remove the old bark from the old apple trees.

4.2 Social sentiment and social taboos

Ladakh has been divided into two districts and the distance between the two districts is 220 km^2. Leh is dominated by Buddhists and Kargil by Muslims. Leh is well known for mixed ethnicity, culture, and traditions, and before performing any operations for the management for codling moth or any field operations, they mostly avoid any management strategies for codling moth during various festivals. As the codling moth is a boring pest, when it enters into the fruit, then it is very difficult to control. Leh district is dominated by the Buddhist population, as the sentiment of killing insects or even pests is considered a sin in major parts of Leh. In Nurla village, we performed all management practices for the management of codling moth. To perform these practices possible in the village, scientific diplomacy was employed to make aware the tribal leaders about the importance of fresh fruit industry, fresh fruit trade, issues pertaining to fruit ban and its prospects on growth and economy. In Kargil district, the sentiment of killing insects is not being observed because no social taboo is related to it but even the population density of codling moth in this region is not below ETL because of various factors (no pruning, training, no proper protection measures and other factors).

4.3 Shops for agriculture implements and products

The dearth of vendors and shops to sell different items (pruning saw, secateurs, foot sprayers, fertilizers, pesticides, pheromone traps and light trap to tribal growers. They are also deficient in to supply the seeds and quality planting material to the growers. The reason for the absence of these implements and planting materials in the private market is discouraged by government policies. These items are partially or fully provided by the government free to the growers under tribal plans. Besides, being a remote area, the season for performing various activities in different ecosystems (horticulture, agriculture) lasts for a few months as the harsh and killing temperatures are prevailing in the region. Tribal growers are unaware and lack knowledge about the importance of orchard management. The tradition or custom of pruning and plant canopy management of apple plants are not being practiced by the tribal growers because of this reason, the height as well as the canopy of the apple plants are not being maintained. The wrong policy and faulty practice of the government is to supply the foot sprayers and insecticides at subsidized rates to the growers is of no use in these situations. Besides, it is beyond the capacity of the foot sprayers to spray the insecticide solution on such untrained and unmanaged apple plants for the management of codling moth in Ladakh. The use of chemical fertilizers and pesticides is not gaining momentum in Ladakh as the government of Ladakh has declared the whole region as an organic belt without any proper guidelines. The ground transport to other states of the Indian union remains cut off for more than six months due to heavy snowfall on Zojilla Pass connecting Ladakh to Jammu and Kashmir. The quality apple plant material/rootstocks which are being produced in bulk in Jammu and Kashmir could not be transported at the proper time in Ladakh. The border of Himachal Pradesh state is connected with Ladakh on another side. This side too remains cut off due to heavy snowfall at different Passes. Therefore, the transportation charges for lifting the planting material or other plant protection products are very costly and not possible to use Cargo flights. In these situations, the government is emphasizing more on to supply the essential commodities to the people of Ladakh, foreign tourists, military, and paramilitary persons during winter. The long hostile borders with China and Pakistan guarded by these troops made the essential commodities more important for their survival and to meet their needs. To summarize it, and the priority of the policymakers is the supply of essential commodities for the survival of the Ladakh populace and military establishments either stored before the winter months or airlifted by cargo flights such as meat, vegetables, and other needs.

4.4 Policies for lifting the ban and way forward

The policies for lifting the ban on all fresh fruits from Ladakh and the ways and methods with the timeline is needed to monitor the population densities of the codling moth and subsequently to reduce the fruit damage on all host trees which harbor pest populations. The scientific advisors and experts are not being consulted on how to move forward for lifting the ban on all fresh fruits from Ladakh. It is not out of place to mention that most of the areas of Ladakh are bordered by Pakistan. Insects have wings and these adults can fly and cross the borders easily (trans-boundary movement). How to reduce such incursions by the codling moth or monitor the population is lacking or both the countries should adopt strict management guidelines for codling moth management, as the hosts for codling moth survival are available on both sides. Recently, the new incursions of browntail moth in Ladakh created havoc on the horticulture crops as this invasive pest is neither reported from India and nor from other bordering states/countries [1]. The flow of tourists without any plant bio-security guidelines are not in place in India or Ladakh. Ladakh is having a very important place with unique biodiversity in the Trans Himalayan region and may be a threat to insect and plant biodiversity due to the introduction of invasive pests (browntail moth, apricot aphid, poplar leaf miner, fruit flies, and locusts) from the last five years [1, 41]. Currently, the unavailability of plant protection measures for this single pest (codling moth) on apple, restricts the movement of all fresh fruits from Ladakh to other states of the Indian Union. To boost fresh fruit trade, it is necessary for the policy makers to adopt and consult codling moth experts and researchers for the management of codling moth in Ladakh. Plant bio-security for the endangered areas which has a monopoly in producing walnuts, apple and other temperature fruits that contribute a major share to the national economy of about 6000 cr and providing employability to a large population of India and also reduce the inflow of these fruits in India to meet out the demand and supply of consumers throughout the year.

4.5 Scientific interventions and awareness campaigns

Timing and use of insecticidal applications are very important for huge infested areas to bring down the population of codling moth below economic threshold levels (ETL). Biofix and degree-day calculations are essential for detecting the first emergence of codling moth adults from the overwintered larvae, monitor the adult activities and egg-laying periods of different generations of codling moth. Though, All India Coordinated Research Project on biological control (AICRP) tried mass production of egg parasitoids viz.; Trichogramma embroyphagum and Trichogramma cacociae could not survive in the cold arid climate of Ladakh. They advocated the use of biocontrol agents and insecticides for the codling moth from the last one decade but desirable results and adoption rate of this technology could not be established as both biocontrol agents and insecticides are antagonistic to each other. The various components of IPM viz., insecticides, botanicals, egg parasitoids, burlapping to target the overwintering population, pheromone baited traps have not been evaluated individually to determine their efficacy against codling moth. Such field trials were laid in small pockets and the benefits of such technologies could not deliver desirable results. The application of these technologies never coincide with the different stage of the target organisms

4.6 Climatic conditions

The climate of Ladakh is cold arid and annually 10 cm of rainfall has been recorded. The huge temperature variations from -30^0 C in winter and $+ 40^0$ C in summers are very challenging for raising plants in such harsh climatic conditions. The gusty winds blowing during summers cause more evapotranspiration from plants and also from the soil surface. Besides, the scarcity of water resources needed for apple plants is not being timed or applied as per the scientific requirements for these plantations. The high tourist flow from the last decade in Ladakh prompted the urban population to establish a maximum number of bore wells in hotels, restaurants, and guest houses for the basic requirement for their comfortable stay and other basic needs. It has been felt that such activities are a threat to the environment and surroundings of Ladakh as the scarcity of water has been felt from the already water-starved region of India. Consequently, the rural people are not very interested to perform various horticultural operations needed from time to time but are being hired by the hotel owners for the maintenance and care for the national and international tourists. A good number of the rural population are fully satisfied to act as guides and transportation facilitators for tourists rather than in the horticulture sector as the avenues for the growth of the horticulture sector are still not opened up due to the ban on all fresh fruits from Ladakh. The scarcity of skilled and semi-skilled persons required to perform various horticulture operations are not being performed. Also, a good number of the local population of Ladakh are migrating to other parts of India to escape from these harsh climatic conditions for more than six months in winter, The big winters are prevailing in the region, and the temperature from October goes beyond sub-zero temperatures till the end of March. Therefore, nurturing and caring for these apple plants and planting material is being left at the mercy of nature.

4.7 Removal of old apple plants and establishing new orchards

More than 30–50% of apple plants in Ladakh are kept untrained and the limbs or branches of apple trees are more robust. The use of burlapping on the main stem to intercept the overwintered larvae of the codling moth is not providing satisfactory results. Our survey team observed that the overwintered larvae of codling moth even overwintered on the adjacent limbs of apple trees. To target, these larvae on these branches could not be achieved because it is very difficult for the person to burlap the whole branches or limbs which harbor these larvae. Mostly these apple trees have developed dead bark, under which it is easier for the codling moth larvae to overwinter during winters. In these circumstances, such apple plants are to be removed and burnt, because such plants have a long history of codling moth infestation. Such apple plants have developed more vegetative growth and to train these plants at this stage is very difficult. These apple plants are so close to each other and it is very difficult for sun rays to reach the ground surface or in the center of these plants which is necessary for growth and the development of quality apple fruits. There is a need to establish new apple orchards on scientific lines by planting elite apple cultivars with deep root stocks which develop fruiting with two years and the quality and quantity of these apple varieties are more than the traditional varieties with fewer inputs with maximum returns.

5. Pheromone dispensers for the management of codling moth in Ladakh and way forward in India

A lot of research work has been carried out using pheromones and mating disruption for the management of codling moth in developed countries [42–44]. Mass trapping is a method to capture enough male insects to get a significant reduction in fruit damage against various pests of economic importance [1]. In codling moth, capturing both partners (male and female) are being targeted using different pheromone traps/dispensers to reduce the high population densities of codling moth [27, 45, 46]. There is scanty information about the number of traps recommended on area basis to target both males and females followed by pheromone dispensers in the world at high population densities. Pheromone baited traps are also being used for monitoring the adult population densities [15, 45, 47]. Mass trapping of codling moth adults has been done in some pockets of Ladakh and trapped a significant number of male adults and also monitoring the adult population to establish the biofix, actually when the adults are flying after winters in Ladakh [15, 26].

6. Demonstration and dissemination of pheromone technology in Ladakh

From 2012 to 2018, demonstrations and dissemination of pheromone technology were conducted in few hamlets of Ladakh during 2012–2018 to observe the efficacy of these treatments as compared to control plots of apple orchards, having a history of codling moth infestation (**Table 3**). All apple plants were installed with a single pheromone trap and the lures were changed three times till harvesting of fruits and pheromone dispensers were installed just after petal fall once in a fruit season. The treatments include, two insecticide applications were sprayed, one at the fruitlet stage and another after thirty days after the first insecticidal spray. The apple plots in Mangbore, Poyeen, and Shilikchey villages were selected for the dissemination and demonstration of pheromone traps and pheromone dispensers and insecticides against codling moth infestation. The treatment with pheromone dispensers +pheromone traps + two insecticidal applications observed an 85% reduction in fruit damage. While dispensers alone recorded 55% fruit damage. While in those treatments, where traps + dispensers were installed, the fruit damage of 69.00% was recorded. All the treatments were compared with control plots. The control plots were two kilometers away from the treated plots.

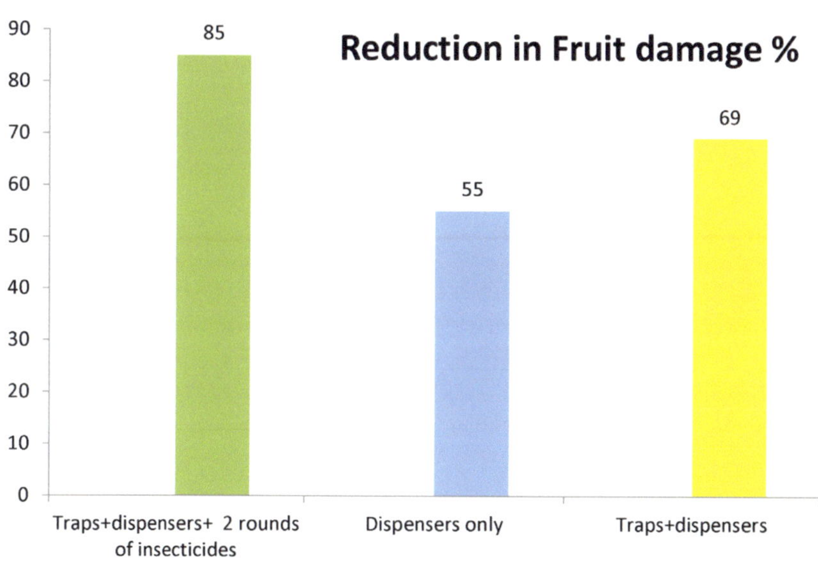

Table 3.
The demonstration of pheromone traps + dispensers in combination and alone for the management of codling moth in Ladakh in selected villages from 2012–2018.

7. Pheromone and mating dispenser technology save water in Ladakh and also in world

After perusing a lot of literature across the globe for the management of the codling moth and other pests using pheromone dispensers and mating dispenser technologies [27, 34]. It has not been reflected that pheromone traps and various mating dispenser technologies can save water. Ladakh is a cold arid region and the scarcity of water is already there in the region. The apple orchards in the different hamlets of Ladakh are without road connectivity and have to travel by foot through small hill slopes between the apple blocks. The availability or the connectivity of water in these apple blocks is not being observed due to natural climatic conditions prevailing in the region. Though for the survival of apple plants, watering for these plants is being carried out by melting of snow on the naked mountains during summer months which precipitates from these hills to lower reaches. Such water is being diverted by the orchardists of Ladakh during the summer months for the survival of these plants. Keeping in view the water scarcity and the factors already detailed above in the Ladakh region. The experts on codling moth in India are emphasizing more on pheromone dispensers and pheromone baited traps for the management of codling moth in Ladakh. The proven technology of pheromone dispensers are easy to apply, no health hazards, easy in application, easy to teach farmers and can be easily adopted by these tribal growers. On average, 15 liters of spray fluid is required for spraying a single apple plant till runoff. Three hundred apple plants are planted

Figure 4.
Pheromone traps and pheromone dispenser technology save water.

in one hectare. As per the estimated (**Figure 4**) of National Horticulture Boards of India, the area under apple cultivation is about 598 hectares. It has been estimated that 4500 liters of water can be saved by using pheromone traps and pheromone dispenser technology for the management of codling moth in Ladakh in a one-hectare area. The application of pheromone traps and pheromone dispensers in Ladakh could save 21.96 Lakh liters of water (total under apple x water requirement for one ha). The pheromone dispenser technology should be adopted in those regions where water scarcity and crisis is more for the management of various insect pests across the globe and environmentally friendly solutions and sustainable agriculture.

To the best of my knowledge, pheromone dispensers are still not registered in India but are used to control various insect pests in the world to their management [48] Therefore, it is important to include pheromone dispensers in India for the management of the codling moth and other insect pests [26] to have safe fruit production without pesticide residues, less fruit damage, safe to the environment and other health issues. Also, this paper may act as a base for policymakers, political advisors on horticulture, researchers, organic growers, and agricultural advisors/commentators of India to use non-chemical management for codling moth [47] and other important pests of India [49] for sustainable and safe agriculture in Ladakh. Pheromone dispenser technology has a huge potential for commercial marketing in India. Besides, reduction of insecticidal usage, reduction of fruit damage, and new employment opportunities in India and Ladakh and subsequently as a safe platform to lift the ban on all fresh fruits from Ladakh.

Author details

Barkat Hussain*, Faizaan Ahmad, Ejaz Ahmad, Wasim Yousuf and Mohd Mehdi
Division of Entomology, SKUAST-K, Shalimar Campus, J&K, India

*Address all correspondence to: bhatbari@rediffamil.com

References

[1] Hussain B, Sivakumar G, Kannan M, War AR, Ballal CR. First record of a nucleopolyhedrovirus infecting brown-tail moth larvae, *Euproctis chrysorrhoea* (L.) (Lepidoptera: Lymantriidae) in India. Egyptian Journal of Biological Pest Control. 2019 Dec 1;29(1):11.

[2] Sharma BR. The special position of Jammu and Kashmir in the Indian constitution. The Indian Journal of Political Science. 1958 Jul 1;19(3):282-290.

[3] Shah IA, Songara M. Production and Marketing Problems of Apple Fruit Growers in Jammu and Kashmir: A Critical Study. MANTHAN: Journal of Commerce and Management. 2019;6(2):57-69.

[4] Vogel B, Field J. (Re) constructing borders through the governance of tourism and trade in Ladakh, India. *Political Geography*. 2020 Oct 1;82:102226.

[5] Rehman MU, Hussain B, Mir MM, Angmo T, Parray E, Zubair M. Low Productivity of Fruits, Its Implications and Combating Strategies in Cold Arid Eco-region of Ladakh (J&K). Current Journal of Applied Science and Technology. 2020 Apr 3:122-128.

[6] Malik, R. A., Punjabi, A. A. and Bhat, A. A. 1972. Survey study of insect and non-insect pests in Kashmir. Horticulturist **13**(3): 29-44.

[7] Wadhi, S. R. and Sethi, G. Wadhi SR, Sethi GS. Eradication of codling moth-a suggestion. Journal of Nuclear Agriculture and Biology. 1975;4(1):18-19.

[8] D O'Rourke A. The world apple market. Routledge; 2018 Dec 19.Source?

[9] Ali J. Analysis of Prices and Arrivals of Apple Fruit in *Narwal Market of Jammu*. Economic Affairs. 2018 Mar 3;63(1):295356.

[10] Shel'Deshova GG. Ecological factors determining distribution of the codling moth Laspeyresia pomonella L.(Lepidoptera, Tortricidae) in the northern and southern hemispheres. *Entomol.* Rev. 1967;46:349-361.

[11] Janjua NA. Codling moth in Afghanistan. Current Science. 1938 Sep 1;7(3):115-116.

[12] Pruthi HS. The distribution, status and biology of codling moth (Cydia pomonella L.) in Baluchistan with notes on some other insects infesting apple. *Indian J. Agr.* Sci. *1938;9:499-547.*

[13] Janjua, N. A., Mustafa, A. M. and Samual, C. K. (1943). On the biology and control of codling moth *Cydia pomonella* L., in Baluchistan. *Indian J.* Agri. Sci., **13**: 112-128.

[14] Pawar AD, Tuhan NC, Balsubramanian S, Parry M. Distribution, damage and biology of codling moth, Cydia pomonella (L). Ind. J. Plant Protection. 1981;10:111-114.

[15] Zaki, F. A. 1999. Incidence and biology of codling moth, *Cydia pomonella* L., in Ladakh (Jammu and Kashmir). Appli. Biol. Res.**1**: 75-78.

[16] Ahmad R, Hussain, B, Ahmad, T. Fresh and dry fruit production in Himalayan Kashmir, Sub-Himalayan Jammu and Trans –Himalayan Ladakh, India. https://doi.org/10.1016/j.heliyon.2020.e05835

[17] Hussain B, Ahmad B, Bilal S. Monitoring and mass trapping of the codling moth, Cydia pomonella, by the use of pheromone baited traps in Kargil, Ladakh, India. *International Journal of Fruit Science*. 2015 Jan 2;15(1):1-9.

[18] Hussain B, War AR, Ganie SA, Bilal S. Monitoring and testing different doses of disparlure for Indian gypsy moth, Lymantria obfuscata, in a temperate region of India (Kashmir Valley). Acta Phytopathologica et Entomologica Hungarica. 2015 Jun;50(1):85-92.

[19] Wilson L, Barnett W. Degree-days: an aid in crop and pest management. California Agriculture. 1983 Jan 1;37(1): 4-7.

[20] Knight AL. Adjusting the phenology model of codling moth (Lepidoptera: Tortricidae) in Washington state apple orchards. Environmental entomology. 2014 Nov 3;36(6):1485-1493.

[21] Epstein DL, Zack RS, Brunner JF, Gut L, Brown JJ. Effects of broad-spectrum insecticides on epigeal arthropod biodiversity in Pacific Northwest apple orchards. Environmental Entomology. 2000 Apr 1;29(2):340-348..

[22] Varela LG, Welter SC, Jones VP, Brunner JF, Riedl H. Monitoring and characterization of insecticide resistance Codling moth (Lepidoptera: Tortricidae) in four Western States. Journal of Economic Entomology. 1993 Feb 1;86(1):1-0.

[23] Mota-Sanchez D, Wise JC, Poppen RV, Gut LJ, Hollingworth RM. Resistance of codling moth, Cydia pomonella (L.)(Lepidoptera: Tortricidae), larvae in Michigan to insecticides with different modes of action and the impact on field residual activity. Pest Management Science: formerly *Pesticide Science.* 2008 Sep;64(9):881-890..

[24] Miller JR, McGhee PS, Siegert PY, Adams CG, Huang J, Grieshop MJ, Gut LJ. General principles of attraction and competitive attraction as revealed by large-cage studies of moths responding to sex pheromone. Proceedings of the National Academy of Sciences. 2010 Jan 5;107(1):22-27.

[25] Knight AL. Managing codling moth (Lepidoptera: Tortricidae) with an internal grid of either aerosol puffers or dispenser clusters plus border applications of individual dispensers. *Journal of the Entomological Society of British* Columbia. 2004;101:69-78.

[26] Mir, W. Y. 2019. Use of Pheromone dispensers in the management of Codling moth in apple and Cucumber fruit flies. Thesis submitted to SKUAST-K in partial fulfilment of the requirement for the award of the M. Sc degree to faculty of Horticulture, 01-47 pages.

[27] Witzgall P, Stelinski L, Gut L, Thomson D. Codling moth management and chemical ecology. Annu. Rev. Entomol.. 2008 Jan 7;53:503-522.

[28] Grant J, Pickel C, Van Steenwyk R, Welter S. Management of codling moth using sprayable microencapsulated pheromones with selective insecticides. Walnut Research Reports. 2004;19:5-206.

[29] Roelofs WL, Bartell RJ, Hill AS, Cardé RT, Waters LH. Codling moth sex attractant—field trials with geometrical isomers. Journal of economic entomology. 1972 Oct 1;65(5):1276-1277.

[30] Riedl H, Blomefield TL, Giliomee JH. A century of codling moth control in South Africa: II. Current and future status of codling moth management. *Journal of the Southern African Society for Horticultural Sciences* (South Africa). 1998.

[31] Bartell RJ, Bellas TE. Evidence for naturally occurring, secondary compounds of the codling moth female sex pheromone. Australian Journal of Entomology. 1981 Aug;20(3):197-199.

[32] Gut L, Wise J, McGhee P, Isaacs R. Pheromone-based control of tree fruit

pests in Michigan, 1998. Where does this reference come from? Source?

[33] El-Sayed A, Unelius RC, Liblikas I, Löfqvist J, Bengtsson M, Witzgall P. Effect of codlemone isomers on codling moth (Lepidoptera: Tortricidae) male attraction. Environmental entomology. 1998 Oct 1;27(5):1250-1254.

[34] Carde RT, Minks AK. Control of moth pests by mating disruption: successes and constraints. Annual review of entomology. 1995 Jan 1;40(1):559-585.

[35] Thomson D, Brunner J, Gut L, Judd G, Knight A. Ten years implementing codling moth mating disruption in the orchards of Washington and British Columbia: starting right and managing for success!. IOBC wprs Bulletin. 2001;24(2):23-30.

[36] Ganie SA, Khan ZH, Ahangar RA, Bhat HA, Hussain B, Liu T. Population dynamics, distribution, and species diversity of fruit flies on cucurbits in Kashmir Valley, India. *Journal of Insect Science*. 2013 Jan 1;13(1).

[37] Vargas RI, Stark JD, Hertlein M, Mafra Neto A, Coler R, Piñero JC. Evaluation of SPLAT with spinosad and methyl eugenol or cue-lure for "attract-and-kill" of oriental and melon fruit flies (Diptera: Tephritidae) in Hawaii. Journal of economic entomology. 2008 Jun 1;101(3):759-768..

[38] Byers JA. Simulation of mating disruption and mass trapping with competitive attraction and camouflage. Environmental entomology. 2014 Nov 3;36(6):1328-1338.

[39] Pfeiffer DG, Kaakeh W, Killian JC, Lachance MW, Kirsch P. Mating disruption for control of damage by codling moth in Virginia apple orchards. Entomologia experimentalis et applicata. 1993 Apr;67(1):57-64.

[40] Gut LJ, Brunner JF. Pheromone-based management of codling moth (Lepidoptera: Tortricidae) in Washington apple orchards. J. Agric. Entomol. 1998 Oct 1;15(4):387-405.

[41] Hussain B, Akbar SA, ur Rehman M, Un Nabi S, Ganie SA, Rasheed War A. Report of Drosophila suzukii (Matsumura)(Diptera: Drosophilidae) from the high-altitude and cold arid region of Ladakh, India. EPPO Bulletin.

[42] McGhee PS, Epstein DL, Gut LJ. Quantifying the benefits of areawide pheromone mating disruption programs that target codling moth (Lepidoptera: Tortricidae). American Entomologist. 2011 Apr 1;57(2):94-100.

[43] McDonough LM, George DA, Butt BA, Gamey LN, Stegmeier MC. Field tests of artificial and natural sex pheromones for the codling moth. Journal of Economic Entomology. 1972 Feb 1;65(1):108-109.

[44] Thomson D. Confusion amongst codling moth fellows continues: a commercial perspective on the implementation of codling moth mating disruption in North America. IOBC wprs Bulletin. 1997;20:57-64.

[45] Knight A, Light D, Chebny V. Evaluating dispensers loaded with codlemone and pear ester for disruption of codling moth (Lepidoptera: Tortricidae). Environmental entomology. 2012 Apr 1;41(2):399-406.

[46] Light DM, Knight AL, Henrick CA, Rajapaska D, Lingren B, Dickens JC, Reynolds KM, Buttery RG, Merrill G, Roitman J, Campbell BC. A pear-derived kairomone with pheromonal potency that attracts male and female codling moth, Cydia pomonella (L.). Naturwissenschaften. 2001 Aug 1;88(8):333-338.

[47] Huang J, Gut LJ, Miller JR. Codling moth, Cydia pomonella, captures in

monitoring traps as influenced by proximately to competing female-like-vs. high-releasing pheromone point sources. Journal of insect behavior. 2013 Sep 1;26(5):660-666.

[48] Stelinski LL, Gut LJ, Mallinger RE, Epstein D, Reed TP, Miller JR. Small plot trials documenting effective mating disruption of oriental fruit moth by using high densities of wax-drop pheromone dispensers. Journal of Economic Entomology. 2005 Aug 1;98(4):1267-1274.

[49] Hussain, B., Masoodi, K. Z., War, A. R., Hakak, A. S., Ahmad, N., & Masoodi, T. (2020). Occurrence of granulovirus infecting Cydia pomonella in high altitude cold arid region of India. Virus Disease, *31*(4), 517-525

Chapter 4

Bioactive Secondary Metabolites of Wild *Antheraea mylitta* Silkworm Cocoons

Sayed Iqbal Ahamad, Kari Neetha and Shyam Kumar Vootla

Abstract

The wild silkworm *Antheraea mylitta* is grown and cultivated in several parts of India ranging from Bihar to West Bengal and several parts of Telangana. The wild silkworm rearing has been a source of income for the tribal populations who rely on it as income source; the intervention of government agencies has increased the cultivation. Our research involves understanding the secondary metabolites in the silkworm Cocoons and elucidating how the pupa survives the harsh environment during pupal diapause of the insect. We have realized the role of insect repellent compounds and other metabolites and their interaction with the insect. Wild silkworm Cocoons are the specialized natural structures constructed by *Antheraea mylitta* silkworms. They are the protein composites of sericin and fibroin as a structural material. The silkworm cocoons are presumed to be evolved structures through the course of evolution over millions of years. This chapter focuses on Biophysical analysis of chemical compounds, proteins and other secondary metabolites traced in the Wild *Antheraea mylitta* Tasar cocoons which are predicted to be the key factors to achieve the unique structural and chemical barriers to protect the pupa within the cocoons.

Keywords: *Antheraea mylitta*, Bioactive compounds, Metabolites, Sericin, Tasar Silkworms

1. Introduction

In wild silkworms host plant specificity is achieved due to the co-evolution of host plants and their monophagous or oligophagous specific herbivorous insect's leads to the accretion of host plant derived allelochemicals in the specific insect cocoons. The plant derived chemicals, play a vital role in the life cycle of the respective phytophagous insects. These bioactive compounds affect the growth, survival, fecundity including behavior of the insects. The economically significant insect cocoons of Tasar silkworm also revealed for biological functions by their secondary metabolites like saponins, flavonoids, terpenoids, tannins and phytosterols sequestered from plant into the larvae to cocoons [1]. The feeding habit and the growing conditions of the silkworms directly influence the chemical composition of the cocoons and the phytochemicals from the host plants of silkworms might be sequestered to cocoons [2]. The secondary metabolites consumed by the silkworms from the host plants are sequestered with silk proteins and play significant role in cocoon formation [3]. In the cocoons of mulberry silkworms three flavonoid 5-glucosides and many other flavonoids of host plant were identified in the sericin layer in yellow-green cocoon of the Sasammayu silkworms. These flavonoids from silkworm cocoons are proved effective for free radical scavenging, antioxidation, inhibition of hydrolytic and oxidative enzymes, and anti-inflammatory action [4] Recently along with mulberry silkworms, the wild non-mulberry silkworms also emerged as commercially significant in textile industry [5] Hence, in present study we focused on the extraction of non-protein active chemical compounds *Antheraea mylitta* cocoons qualitatively and validated by using Fourier Transform-Infrared spectroscopy (FT-IR) and Gas chromatography–Mass spectrometry (GC–MS). The biological activity of the compounds was screened by compared; the phytochemicals from the cocoons were further confirmed in their respective host plants from the earlier reports to elucidate the arthropod-host plant interactions to predict the sequestration of allelochemicals from host plants to the insect cocoons.

As compared to domesticated mulberry silkworm cocoons, the wild silkworms and their cocoons show slightly different combination of morphological, chemical properties and are adapted to cope with harsh natural conditions. By considering this, we selected commercially exploited silkworms of domesticated and wild silkworms to screen the active chemical components from their cocoons and their sequestration patterns from feeding plants to the cocoons by qualitative methods.

The cocoons of most insect larvae are complex structures potentially serving various synchronized functions. The silkworm cocoon is generally presumed to provide the protection to inactive pupa against predation, biodegradation, dehydration etc. Among the most extensively studied cocoons *Bombyx mori* cocoons take the lead. Silk fabric has been valued in numerous cultures for many millennia [6]. As per commercial production of silk *Bombyx mori* silk has been comprehensively investigated. As representative of many holometabolous insects, the silkworm life cycle start as a larva, passing through five larval instars and after the completion of fifth larval instar, the reduction of the juvenile hormone permit the neurosecretory hormone ecdysone to initiate metamorphosis and activate the initiation of the prepupal stage. The prepupa locates the suitable place for cocoon formation, then start to spin the cocoons. The silkworms spun the cocoons around the constricted body of the pupa which uses silk strands secreted from labial glands [7]. Silk strands themselves are polypeptide polymers composed of multiple components-microfilaments of insoluble proteins (fibroin), covered with a soluble adhesive protein (sericin) that provides structural support for the cocoon [8]. Other minor components include small proteins, lipids, and carbohydrates [9].

The wild sericigenous Indian tropical Tasar Silkworm *Antheraea mylitta* is a polyphagous insect feeding on variety of leaves; it has rich genetic resources of forty four races acclimatized to diverse ecological zones. In the course of evolution it has been evolved with many advanced qualities such as disease resistance, silk quality, fecundity and tolerance to various environmental conditions (**Figure 1**). The wild Tasar cocoons are exposed to various biotic and abiotic environmental conditions, hence the quality of the silk, silk proteins and other secondary metabolites of the cocoons are differing than of mulberry silkworm cocoons [10, 11]. Comparatively the domesticated *Bombyx mori* cocoons are soft and delicate, only the hot water treatment swells and partly dissolves the sericin gum, which coats and cements the fibroin filaments together inthe cocoon. But Wild silkworm species including Tasar silkworms are heavily mineralized with calcium oxalates (**Figure 2**) [12]. In addition to this wild cocoon are additionally stabilized by oxidative phenolic tanning, dityrosine cross-linking, and tannins derived from the caterpillar's food

Figure 1.
Antheraea mylitta *Silkworm.*

Figure 2.
Scanning Electron Micrograph of calcium oxalate crystals on Antheraea mylitta *cocoons.*

Figure 3.
Antheraea mylitta *Silkworm cocoon spinning.*

Figure 4.
Antheraea mylitta *cocoons collected from forest.*

plant. Mineralization is probably more important than the tanning in making Wild Silk cocoons difficult to soften and reel (**Figure 3**) [13]. The remarkable contrast in the composition of cocoons, the methods of obtaining silk from the cocoons of wild Silkworm with those domesticated *Bombyx mori* makes it complicated. The most important difference lay in the mineralization that very much common in wild Silkworms but absent in *Bombyx mori* and the difference arise from the gluing together of the fibers in mineralized matrix of wild silk cocoons made them moisture resistant (**Figure 4**) [14].

2. Bioactive silk proteins

The cocoons of the mulberry silkworm are composed of two types of proteins: fibroins and sericins. The fibroin is the core protein constitutes 70% of the cocoon and is a hydrophobic glycoprotein secreted from the posterior part of the silk gland. The expression of fibroin protein and P25 genes are transcriptionally regulated during larval development in both *Bombyx mori* and *Antheraea mylitta*. The fibroin protein is semi-crystalline in nature, showing of two phases, i. e greatly crystalline β-sheeted phase and a lesser or non-crystalline phase. The hydrophilic sericin proteins constitute about 20–30% of the cocoon which is hot water-soluble glycoproteins and hold the silk fibers together form the ecologically stable sericin-fibroin composite cocoon structure. The glue like sericin protein biosynthesized and secreted in the middle region of the silk gland. It comprises diverse polypeptides ranging from 24 to 400 kDa with high serine content (40%) with considerable amount of glycine (16%). There are three major polypeptides of sericin have been isolated from the silkworm cocoons with the molecular weights ranges from 150, 250 and 400 kDa. The sericin residues are partially unfolded with 35% β-sheet and 63% random coil, without α-helical structures. In addition to these major proteins, the low molecular weight hydrophilic proteins are also reported in the cocoons. The seroin protein is the product of a discrete gene that is expressed exclusively in the middle and the posterior part of the silk glands [15–20].

Indian tropical wild *Antheraea mylitta* silkworms are of Saturniidae respectively. Silkworms produce delicate twin thread of silk protein fibroin, which is coated by glue like hydrophilic sericin protein (**Figure 5**). During pupation silkworms spin the cocoons to protect the inactive pupae. The silk proteins are synthesized by silk gland cells and stored in the lumen of the glands. The sericin protein is biosynthesized in the middle silk gland of the mature silkworm larvae, which constitutes 25–30% of silk proteins [21]. It is a water-soluble globular protein family whose molecular mass ranges from 10 to 310 kDa [22]. Naturally sericin is responsible for adhere both the fibroin filaments to maintain the structural integrity of the cocoon. The cocoon of *Antheraea mylitta* has three major fractions of sericin of which the lower fraction is around 70 kDa, the middle fraction is approximately 200 kDa, and the higher fraction is more than 200 kDa. The peduncle of Tasar silkworm *Antheraea mylitta* has a 200-kDa sericin protein, possesses serine, glutamic acid, glycine, tyrosine and threonine as predominant amino acid residues. The Serine (~39%) is the principal amino acid. The *Antheraea mylitta* silk sericin is biochemically distinct from *Bombyx mori* having lower percentages of serine and tyrosine. During degumming process of silk textile industry, sericin is removed as waste from fibroin to make silk fibers more lustrous, soft, smooth, white, and dye able. The global discarded sericin constitutes approximately 50,000 tons out of the 1 million tons of fresh cocoons annually [15]. Silk sericin of *Bombyx mori* is one of the most researched proteins. Presently sericin can be used in food, cosmetics, pharmaceutical products and the preparation of biomaterials. The sericin proved as antioxidant, anticoagulant and anti-wrinkle agent. It is also reported to suppress tumor growth and to reduce oxidative stress [23–25].

The domesticated *Bombyx mori* silk sericin contains 18 amino acids including polar amino acids such as 32% serine and 17% aspartic acid gives higher hydrophilic property and processing ability [26]. In contrast to this Tasar silk sericin contains 19% serine contents. The mulberry and Tasar sericins are biochemically distinct due to differences in their amino acid compositions, leading to differences in the immunological responses. Being non-domesticated and wild, Tasar cocoons are more resistant towards environmental stresses such as heat and drying, the sericin coat may contributed for toughness and resistance properties [27].

The silk of domesticated and wild silkworms has a core shell type structure, it is composed of a complex of 3 proteinaceous components: a large heavy chain fibroin (350 kDa) that is linked to a light chain fibroin (25 kDa) by disulfide bonds and another glycoprotein P25 protein (30 kDa) are linked with non-covalent

Figure 5.
Scanning Electron Micrograph of Antheraea mylitta *cocoon surface.*

hydrophobic interactions [28]. The molar ratios of Heavy chain, Light chain and P25 are 6:6:1. The heavy chain is hydrophobic in nature and makes crystalline features to the silk fiber, but the Light chain is more hydrophilic and comparatively elastic. The P25 protein is supposed to play a crucial role to maintain the integrity of the complex [29]. Prior to silk fiber formation, the solution of all the three silk proteins secreted from silk glands assembling into double filaments that come out from an exit tube in its spinneret and dry after exposure to air. Consequently core contains anisotropic β-sheet-rich nanocrystals are loosely aligned with the fiber axis and dispersed in an unstructured matrix [30]. Another pair of silk glands secrete glue-like sericins (serine-rich glycoproteins) that coat the fibroin filaments for the cohesion of the cocoon by sticking the twin filaments together. The silk fibers coated with several other proteins are presumed to protect the cocoon against microorganisms and other predators [31, 32].

2.1 Biomedical applications of silk proteins

The silk proteins are biologically versatile molecules in the context of biomedical applications. The silk fibers have been used as sutures for wounds since many centuries, because of its strength, biocompatibility and low immunogenicity [33]. Even the silk fibers spun into yarns and consequently textured via permanent deformation may be used as non-load-bearing spacers in tissue grafts where tissue in-growth is desirable. The cabled yarns have great tunable mechanical properties and therefore potential in load-bearing tissue engineering applications [34]. In addition to these silk foams prepared from fibroin has been used as scaffolds for the attachment and proliferation of fibroblasts *in vitro* condition. The study on biomedical applications of silk protein reveal that the cultured cell colonies were located at the surface of the foam, preferably due to the cell-seeding process due to lacking nutrients inside the foam [35]. Silk-based materials have been used as organic scaffolds for the biomineralization of hydroxyapatite and silica. The silk proteins are considered as potential molecules for a drug release profile and are both reliable and controlled, particularly important in cases where the drugs have undesirable side effects. Silk proteins may find application in drug delivery as drug carriers owing to their biocompatibility and their highly tunable morphologies [36].

Silk protein-based materials have found application as solid supports for potentially expensive enzyme and organometallic catalysts. Silk proteins are capable of forming functional complexes with metal ions. The enzymes can be successfully immobilized by means of covalent linking of the silk protein to the enzyme using conventional cyanogen bromide, azide, diazo or glutaraldehyde methodologies. The aspartate aminotransferase enzyme, calf intestine alkaline phosphatase enzyme and ribonuclease enzymes have been covalently linked to *Bombyx mori* fibroin effectively and were shown their activity efficiently. Many enzymes effectively immobilized through physical entrapment method within the silk films [37–40].

The silk sericin has many biomedical applications as antioxidant, anticancer drug and anticoagulant. The study on macrophage response of silk proteins concludes that silk sericin does not allow inflammatory response when supply in soluble form. But the macrophage activation study of silk sericin reveals that, when attached to fibers induce inflammatory responses [41]. The silk sericin in presence of lipopolysaccharides shows inflammatory reaction by initiating the synthesis of tumor necrosis factor and which is connected with native silk fiber-induced immune responses [42–44]. The explanation could be that coated sericin proteins either provide better adhesion to macrophages or the structural changes of sericin after binding to silk fibers prime the macrophage for consequent stimulation.

The bioconjugation (e.g. polymer-protein) is beneficial because it leads to minimize immunogenicity and improve stability. The biopolymer conjugations with anticancer drug candidates promote tumor targeting efficiency through superior permeability and increase drug retention time. Many investigators have utilized sericin as a natural biopolymer for bioconjugation with various therapeutic proteins, enzymes and polysaccharides. The sericin is the best natural biopolymer which can be conjugated effectively due to the presence of functional surface-active groups (-OH, -COOH, -NH2), which can form covalent linkage with the conjugates. The success of best performance of sericin is due to its hydrophilic nature and low antigenicity and other immunogenic properties and higher half-life period *in vivo* due to filtration by the kidneys to increase retention period [45–49].

The two-dimensional films and three-dimensional matrices of hydrogels and porous scaffolds of sericin proteins have been reported for their better performance. The membranes of sericin are naturally fragile in the dry state. The blending of sericin with water-soluble polymers like polyvinyl alcohol for making films has been investigated. The sericin hydrogels blend with polyvinyl alcohol by irradiation at 40 kGy has been reported. The blended hydrogels show an excellent moisture adsorbing tendency, desorbing and elastic properties with potential applications as soil conditioners to biomaterials for biomedical applications including wound dressings [50, 51].

3. Bioactive Tasar cocoon secondary metabolites

In present research the qualitative analysis of phytochemicals were tested for alkaloids, saponins, steroids, phenols, flavonoids, terpenoids, tannins, fatty acids, carboxylic acids, volatile oils, fixed oils and aldehydes in the methanolic extracts of Tasar cocoons by using standard methods [52] and compiled in **Table 1**. The alkaloids and volatile oils are present in Tasar cocoons whereas terpenoids and tannins are absent. The tests for phenols, fatty acids, carboxylic acids and fixed oils have shown positive reports.

The Tasar cocoon extract contained 14 characteristic GC–MS peaks emerged, which represents various respective chemical components in the extract (**Figure 6**). The Tasar cocoon methanolic extract revealed fourteen characteristic GC–MS peaks which are correlated with FT-IR (**Figure 7**), which represent their respective chemical components in the extract. The chromatogram maximum peak area percentage was observed for 26-Nor-5-cholesten-3.beta-ol-25-one (65%), Oleic acid (9.47%), n-Hexadecanoic acid (7.24%), Stegmasterol (5.93%) and Octadecanoic acid (5.51%). The FT-IR spectral analysis of the compounds of Tasar cocoons are presented in **Table 2** and chemical structures are presented (**Figure 8**). These compounds are reported as potent antimicrobial, anti-inflammatory, anticancer, antioxidant, Hypocholesterolemic in nature respectively [53–59]. Other compounds shown minimum peak area percentage it directly dictates less in concentration. The complete GC–MS analysis and the biological activity of the compounds are presented in **Table 3**.

3.1 4-Methyltridecane and isobutyl alcohol

4-methyltridecane is a branched alkane consisting of tridecane bearing a single methyl substituent at position 4. It has a role as a plant metabolite. It is reported for antioxidant properties [60].

Bioactive compounds	Qualitative tests
Alkaloids	+
Saponins	+
Steroids	+
Phenols	+
Flavonoids	+
Terpenoids	-
Tannins	-
Fatty acids	+
Carboxylic acids	+
Volatile oils	+
Fixed oils	+
Aldehydes	-

Table 1.
Qualitative tests for active compounds from Methanolic extract of Antheraea mylitta *cocoons.*

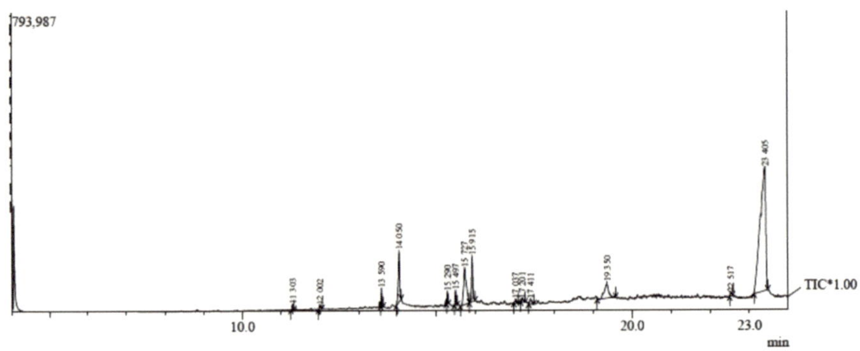

Figure 6.
GC-MS chromatogram of Antheraea mylitta *cocoon extract.*

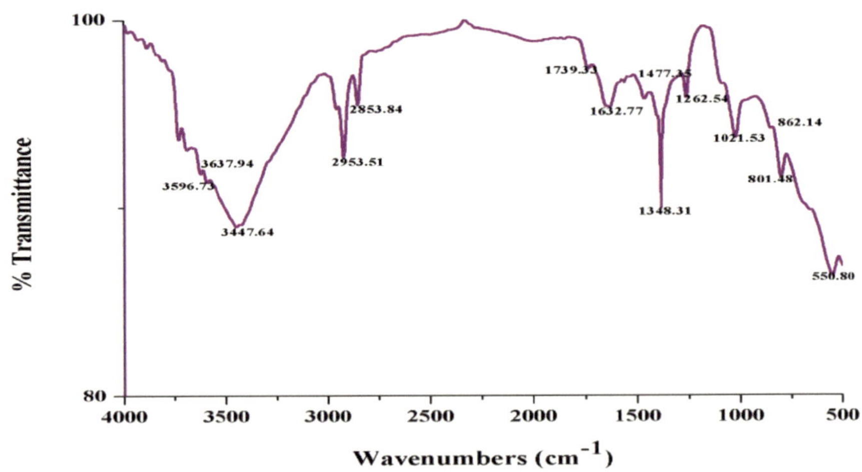

Figure 7.
FT-IR of Antheraea mylitta *cocoon extract.*

FT-IR spectra of Tasar cocoons (Wavenumbers cm-1)	Bonds and Structures	Description
3637.94 3596.73	O-H stretch, free hydroxyl, H-bonded	Alcohols and Phenols
3447.64	N-H stretch	1°, 2° amines, amides
2953.51 2853.84	C-H stretch	Alkanes
1739.33 1632.77	C=O stretch	Amide I
1477.35	C-N Stretching, N-H bending	Amide II
1348.31	N-O symmetric stretch	Nitro compounds
1318.50	C-O Stretch	Alcohols, carboxylic acids, esters, ethers
1262.54 1021.53	C-N Stretching, N-H bending	Amide III (Aliphatic amines)
862.14	O-H bending	Carboxylic acids
801.48	C-Cl Stretch	Alkyl halides
550.80	C-H Stretch	Aromatics

Table 2.
FT-IR spectra of Antheraea mylitta *cocoon extracts.*

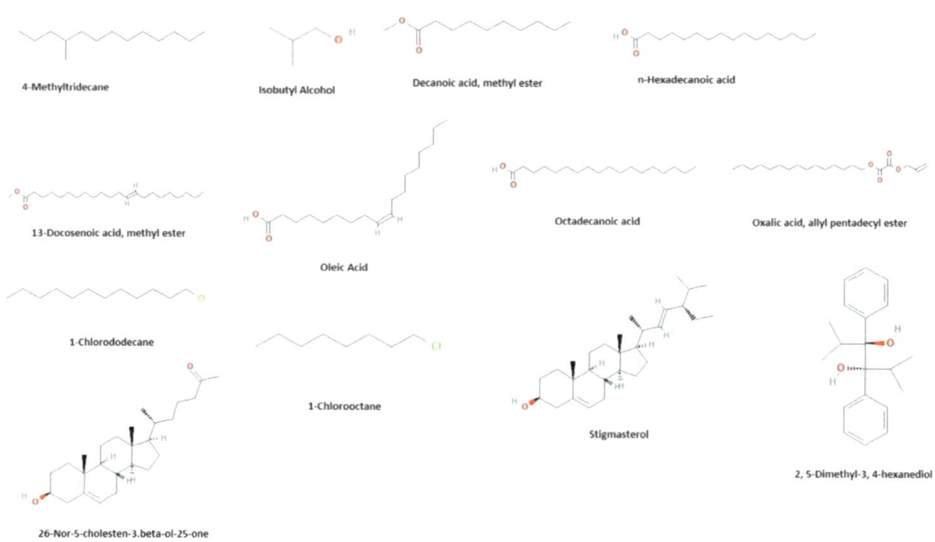

Figure 8.
Chemical structures of Identified Bioactive compounds from Antheraea mylitta *Cocoon Extract.*

Isobutanol is an alkyl alcohol substituted by a methyl group at position 2. In *Saccharomyces cerevisiae* it acts as active metabolite for various physiological functions. This primary alcohol derives from a hydride of an isobutene [61].

3.2 Decanoic acid, methyl ester and N-hexadecanoic acid

Methyl decanoate is a fatty acid methyl ester and a decanoate ester, it is reported for Antioxidant activity, Antibacterial, antiviral, antifungal activity [62].

Name	RT (m)	Area (%)	Structure	Molecular weight	Chemical Nature	Bioactivity
4-Methyltridecane	11.30	0.45	$C_{14}H_{30}$	198	Alkane	Insect predator and Pheromone
Isobutyl alcohol	12.00	0.51	$C_4H_{10}O$	74	Alcohol	Alcohol detoxicant, disinfectant
Decanoic acid, methyl ester	13.59	1.38	$C_{11}H_{22}O_2$	186	Fatty acid ester	Food additive and lubricant, Insecticide
n-Hexadecanoic acid	14.05	7.24	$C_{16}H_{32}O_2$	256	Palmitic acid	Antioxidant, nemeticide and Hypocholesterolemic
13-Docosenoic acid , methyl ester	15.29	0.84	$C_{22}H_{44}O_2$	352	Fatty acid ester	Lubricant and surfactant
Pentanoic acid, 4-methyl, methyl ester	15.49	1.30	$C_7H_{14}O_2$	130	Fatty acid ester	Flavouring agent
Oleic acid	15.72	9.47	$C_{18}H_{34}O_2$	282	Fatty acid	Anticancer, anti-aging and Anti-inflammatory
Octadecanoic acid	15.91	5.51	$C_{18}H_{36}O_2$	284	Fatty acid	Adhesive and sealant
Oxalic acid, allylpentadecyl ester	17.03	0.42	$C_{20}H_{36}O_4$	340	Carboxylic acid ester	-
1-Chlorododecane	17.20	0.54	$C_{12}H_{25}Cl$	204	Alkyl halide	Antitumor, Increase natural killer cell activity
2,5-Dimethyl-3,4-hexanediol	17.41	0.93	$C_8H_{18}O_2$	146	Alcohol	-
Stigmasterol	19.35	5.93	$C_{29}H_{48}O$	412	Steroid	Antioxidant, hypoglycemic and thyroid inhibiting properties, Precursor of progesterone, Antimicrobial, Anticancer, Antiarthritic, Antiasthama, Anti-inflammatory, Diuretic.
1-Chlorooctane	22.51	0.49	$C_8H_{17}Cl$	148	Alkyl halide	Anticancer
26-Nor-5-cholesten-3. beta-ol-25-one	23.40	65.00	$C_{26}H_{42}O_2$	386	Steroid	Antimicrobial, Diuretic, Anti-inflammatory, Anti-asthma

Table 3.
GC-MS Identified Bioactive Secondary Metabolites from Antheraea mylitta *cocoon.*

The hexadecanoic acid is a straight-chain, sixteen-carbon, saturated long-chain fatty acid. N-hexadecanoic acid and it showed significant cytotoxicity against human colorectal carcinoma cells. It is also reported as anti-inflammatory compound [63]. Hexadecanoic acid, methyl ester exhibited antioxidant, hypocholesterolemic, anti-androgenic, hemolytic, alpha reductase inhibitor activities [64]. The major saturated fatty acid hexadecanoic acid has recently been shown to be neutral in its cholesterolaemic effect. The Palm oil rich with hexadecanoic acid, the consumption has been reported to reduce blood cholesterol in comparison with the traditional sources of saturated fats such as coconut oil, dairy and animal fats [65].

3.3 13-Docosenoic acid, methyl ester

13-Docosenoic acid, methyl ester is a fatty acid methyl ester, that is a flavor-active, volatile, and aromatic compound found in cooked commercial shrimp waste. It is a component of biodiesel formed from *Croton megalocarpus* and *Ceibapentandra* oils that contain trierucin. 13(Z)-Docosenoic acid methyl ester has also been used [66].

3.4 Oleic acid

Oleic acid has been reported to have hypocholesterolemic, antioxidant and lubricating activity [67]. Oleic acid is commonly found in diet. It is a monounsaturated fat which on consumption has been linked with decreased lowdensity lipoprotein cholesterol, and possibly increased high-density lipoprotein cholesterol [68].

3.5 Octadecanoic acid and oxalic acid, allylpentadecyl ester

A C18 straight-chain saturated fatty acid component of many animal and vegetable lipids. As well as in the diet, it is used in hardening soaps, softening plastics and in making cosmetics, candles and plastics. It is a stearic acid ester reported for its antioxidant and anti-inflammatory activity [69]. The alcoholic compound of oxalic acid is reported for antimicrobial preservative [70].

3.6 1-Chlorododecane

1-Chlorododecane belonging to the family of organic halogen compounds. It is hard to dissolve in water but can be mixed with alcohol and ether. This chemical is less health hazard substance than short chain alkyl chlorides. It is used as a solvent, as chemical intermediate to make photographic chemicals, pharmaceuticals, organometallic compounds, surfactants [71].

3.7 2, 5-Dimethyl-3,4-hexanediol and stigmasterol

2, 5-Dimethyl-3, 4-hexanediol extracted from *Phormidium autumnale* is reported for antimicrobial activity [72]. Another steroidal compound Stigmasterol reported for Anti-tumor, Cancer preventive, inhibit intestinal cholesterol absorption, antiinflammatory activity [73].

3.8 1-Chlorooctane

The GC–MS analysis of essential oils obtained from the peel of *Citrus reticulata* was confirmed for1-Chlorooctane as an important compound. It has reported for marked antibacterial and antifungal activities, as evidenced by their zones of inhibition. Among the tested microbiology, the oil was very active against *Bacillus subtilis*, *Aspergillusflavus*, *Escherichia coli* and *Staphylococcus aureus* [74].

3.9 26-Nor-5-cholesten-3.beta-ol-25-one

The steroid compound of cholesten reported for Antimicrobial, Diuretic, Anti-inflammatory, anti-asthmain acetone extract of *Cenchrusse tigerus* [75].

The primary host plant for Tasar silkworm is *Terminalia arjuna* (Combretaceae), it has been reported for its antimicrobial properties and to treat cardiovascular disease [76, 77]. In this study similar phytochemical compounds of *Terminalia arjuna* leaves observed in Tasar cocoons but steroids are not observed in the Tasar host plant [78, 79]. The alkaloids, saponins, steroids, flavonoids, terpenoids, tannins, volatile oils and aldehydes are not observed in the cocoon extract. This observation highlights in addition to the phytochemical sequestrations from host plant to silkworm cocoons, even biosynthesis of the other active compounds by the silkworm larvae takes place during spinning of the cocoons. The mechanism of metabolic pathways for the phytochemical sequestrations or the synthesis of the active compounds observed in the mulberry and wild silkworms need further exploration.

4. Conclusion

The silk fibers and insect extracts have been comprehensively used in folk medicines from thousands of years. In Chinese traditional medicines insects and insect-based products are used for various diseases and ailments. The insects and their products have evolved individually over the track of evolution; naturally insects face numerous biotic and abiotic challenges in their life cycles. Due to microorganisms infested habitats they occupied their success in the survival, infinite numbers and diversity specify the presence of extremely effective immune systems produce powerful antimicrobial, cytotoxic compounds for the parasites and other medicinally valuable chemical compounds. The chemical defense strategies of insects have evolved, including odorous repellents to avoid or to kill or inactivate the defending individual predatory organisms including microorganisms. By considering this, we designated the commercially exploited lepidopteran insect wild Tasar silkworm cocoons to screen the active non-protein chemical components [80].

Naturally, silkworm pupa is enclosed within the cocoons during the metamorphosis from pupa to adult. This is the most susceptible stage for the insects because the immobile pupa is not able to respond to biotic and abiotic threats. Therefore, cocoon provides or modifies the microenvironment of the pupa to ensure the optimal conditions for successful pupation and possesses antimicrobial properties. In comparison to wild silkworms, domesticated mulberry silkworms are more privileged due to their domestication and indoor rearing practices. Hence the structural and chemical composition of wild silkworms is strong enough to face threats during their life cycle. The diverse chemical compounds with strong biological activities were identified in the wild silkworm cocoons compared to mulberry silkworms. The comprehensible mechanism of bioactive compound synthesis and their sequestration from specific host plants to the silkworm cocoons and their function for the silkworm physiology is yet to be explained. The insect specific chemical products including antimicrobial compounds, their biosynthesis strategies and the mechanism of action may take part in the discovery of new drug candidates in the field of biomedical science.

The Bioactive components in addition to the Silk proteins from Tasar cocoons were identified by various methods and their biomedical application was compiled. The qualitative analysis of the extracts was performed to validate the active chemical compounds. The conclusion of our results drawn as the phytochemical of the host plants sequestrated to cocoons and the biosynthesis of bioactive compounds by the silkworm larvae during spinning of the cocoons to protect pupae during metamorphosis. But the molecular mechanisms or metabolic pathway for phytochemical sequestration or the biosynthesis of the bioactive compounds in the insects need further research. In comparison to mulberry cocoons, non-mulberry cocoons possess the antimicrobial and insect repellent agents, which are presumed to be involved in the direct protection of wild cocoons by microbial decomposition and other insect predators in wild environmental conditions. We conclude the physico-chemical interactions of the cocoons are responsible to protect the inactive pupae during metamorphosis. Further exploration is required for strategic isolation of cocoon protecting active compounds from mulberry and non-mulberry cocoons for biomedical applications.

Acknowledgements

Prof. V. Shyam Kumar thanks to Department of Science and Technology- Science and Engineering Research Board (DST-SERB), New Delhi. India for funding of this project (SB/EMEQ-154/2013) and Authors are thankful to the University Science Instruments Center (USIC), DST-FIST, Dept of Biotechnology & Microbiology, Karnatak University Dharwad for GC–MS, FT-IR studies.

Conflict of interest

"The authors declare no conflict of interest."

Author details

Sayed Iqbal Ahamad[1], Kari Neetha[2] and Shyam Kumar Vootla[3*]

1 Department of Biotechnology, Khaja Bandanawaz University, Kalaburagi, India

2 Department of Zoology, Karnatak Science College, Dharwad, India

3 Department of Biotechnology and Microbiology, Karnatak University, Dharwad, India

*Address all correspondence to: vootlashyam@gmail.com

References

[1] Bindu PC, Jaisankar P, Hauer F et al (2006) Biological relevance of host plant-derived terpenoid in the cocoons of the tropical Tasar silkworm *Antheraea mylitta*. Biochem Syst Ecol 34(9):698-704

[2] Bharani A, Ganguli A, Mathur LK et al (2002) Efficacy of *Terminalia arjuna* in chronic stable angina: a double-blind, placebo-controlled, crossover study comparing *Terminalia arjuna* with isosorbide mononitrate. Indian Heart J 54(2):170-175

[3] Kumar JP, Mandal BB (2017) Antioxidant potential of mulberry and non-mulberry silk sericin and its implications in biomedicine. Free Radic Biol Med, 108:803-818

[4] Wang HY, Wang YJ, Zhou LX et al (2012) Isolation and bioactivities of a non-sericin component from cocoon shell silk sericin of the silkworm *Bombyx mori*. Food Funct, 3(2):150-158

[5] Arunkumar KP, Tomar A, Daimon T et al (2008) Wild Silk base: an EST database of wild silkmoths. BMC Genomics 9(1):338

[6] Kadolph, Sara J., 2007. Textiles, 10th ed. Pearson Prentice Hall, Upper Saddle River, pp. 76-81.

[7] Asakura, T., Umemura, K., Nakazawa, Y., Hirose, H., Higham, J., Knight, D., 2007. Someobservations on the structure and function of the spinning apparatus in the silkworm *Bombyx mori*. Biomacromolecules 8, 175-181.

[8] Hakimi, O., Knight, D.P., Vollrath, F., Fadgama, P., 2007. Spider and mulberry silkworm silks as compatible biomaterials. Compos, Part B Eng. 38, 324-337.

[9] Gauthier, N., Mandon, N., Renault, S., Bénédet, F., 2004. The *Acrolepiopsis assectella* silk cocoon: kairomonal function and chemical characterization. J. Insect Physiol. 50,1065-1074.

[10] Sayed Iqbal Ahamad M and Shyam KumarVootla.2018, Extraction and Evaluation of Antimicrobial Potential of *Antheraea mylitta* Silk Sericin. Int JRecent Sci Res. 9(1), pp. 23019-23022. DOI: http://dx.doi.org/10.24327/ijrsr.2018.0901.1382

[11] Ahmad, S.E., Kamra, A., Hasnain, S.E. (2004). Fibroin silkproteins from the non-mulberry silkworm *Philosamia ricini* are biochemically and immunochemicallydistinct from those of mulberry silkworm *Bombyx mori*. DNA Cell Biol., 23: 149-154.

[12] Akai, H., & Nagashima, T. (2003). Calcium crystals of cocoon shell from African Gonometa silkmoth (Lasiocampidae). International journal of wild silkmoth & silk, 8, 1-5.

[13] Shamitha, G., & Rao, A. P. (2006). Studies on the filament of Tasar silkworm, Antheraea mylitta D (Andhra local ecorace). Current science, 1667-1671.

[14] Gheysens, T., Collins, A., Raina, S., Vollrath, F., & Knight, D. P. (2011). Demineralization enables reeling of wild silkmoth cocoons. Biomacromolecules, 12(6), 2257-2266.

[15] Couble P, MoineA, GarelA, Prudhomme JC. Developmental variationof a non-fibroin mRNA of *Bombyx mori* silk gland, encoding a lowmolecular silk protein. Dev Biol 1983;97:398-407.

[16] Zhou C, Confalonieri F, Medina N, Zivanovic Y, Esnault C, Yang T, et al. Fine organization of *Bombyx mori* fibroin heavy chain gene. Nucleic Acids Res 2000;28:2413-9.

[17] Kikuchi Y, Mori K, Suzuki S, Yamaguchi K, Mizuno S. Structure of the *Bombyx mori* fibroin light chain-encoding gene: upstreamsequence elements common to the light and heavy chain. Gene1992;110:151-8.

[18] Tanaka K, Kajiyama N, Ishikura K, Waga S, Kikuchi A, Ohtomo K, et al. Determination of the side of disulfide linkage between heavy and light chains of fibroin produced by *Bombyx mori*. Biochim Biophys Acta 1999;1432:92-103.

[19] Mandal, B. B., Ghosh, B., & Kundu, S. C. (2011). Non-mulberry silk sericin/ poly (vinyl alcohol) hydrogel matrices for potential biotechnological applications. International Journal of Biological Macromolecules, 49(2), 125-133. https://doi.org/10.1016/j.ijbiomac.2011.03.015

[20] Datta A, Ghosh AK, Kundu SC. Purification and characterization of fibroin from the tropical saturniid silkworm, *Antheraea mylitta*. Insect Biochem Mol Biol 2001;31:1013-8.

[21] Dash, R., Ghosh, S. K., Kaplan, D. L., & Kundu, S. C. (2007). Purification and biochemical characterization of a 70kDa sericin from tropical Tasar silkworm, *Antheraea mylitta*. Comparative Biochemistry and Physiology - B Biochemistry and Molecular Biology, 147(1), 129-134. https://doi.org/10.1016/j.cbpb.2007.01.009

[22] Sehnal, F., & Žurovec, M. (2004). Construction of silk fiber core in Lepidoptera. Biomacromolecules, 5(3), 666-674.

[23] Gupta, D., Agrawal, A., & Rangi, A. (2014). Extraction and characterization of silk sericin. Indian Journal of Fibre and Textile Research, 39(4), 364-372.

[24] Zhou CZ, Confalonieri F, Jacquet M, Perasso R, Li ZG, Janin J. Proteins –Structure, Function and Genetics 2001;44(2):119-22.

[25] Marsh, r. E., Corey, r. B., & Pauling, l. (1956). Die struktur von Tussah-seidenotizuber die struktur von beta-poly-l-alanin. In angewand techemie-international edition (vol. 68, no. 6, pp. 218-219). Muhlenstrasse 33-34, d-13187 Berlin, Germany: wiley-v chverlaggmbh.

[26] Altman, G. H., Diaz, F., Jakuba, C., Calabro, T., Horan, R. L., Chen, J.& Kaplan, D. L. (2003). Silk-based biomaterials. Biomaterials, 24(3), 401-416.

[27] Kaur, N., Shafiq, N., Negi, H., Pandey, A., Reddy, S., Kaur, H., & Malhotra, S. (2014). *Terminalia arjuna* in chronic stable angina: Systematic review and meta-analysis. Cardiology research and practice, 2014.

[28] Smritikana Biswas, PranoyMitra. (2017). Comparative studies on the antimicrobial activity of *Terminalia arjuna* and *Aloe vera* against community associated drug resistant *Staphylococcus aureus* in pus from carbuncles of adults. Journal of Innovations in Pharmaceutical and Biological Sciences, 4 (4), 153-157

[29] Desai, D., &Chanda, S. (2014). Pharmacognostic Study and Physicochemical Analysis of Leaves of *Terminalia arjuna*. Pharmacognosy Journal, 6(6), 15-19.

[30] Ahamad, M. S. I., Neetha, K., &Vootla, S. K. (2020). Non-protein Chemical Compounds from Lepidopteran Insect Cocoons. In Natural Materials and Products from Insects: Chemistry and Applications (pp. 137-156). Springer, Cham.

[31] Aramwit, P., Siritientong, T., Srichana, T., 2012. Potential applications of silk sericin, a natural protein from textile industry by products. Waste Manag. Res. 30, 217-224.

[32] Akai, H. (1997). Anti-bacterial function of natural silk materials. Int. J. Wild Silkmoth& Silk, 3, 79-81.

[33] Vollrath, F., Barth, P., Basedow, A., Engström, W., & List, H. (2002). Local tolerance to spider silks and protein polymers in vivo. In vivo (Athens, Greece), 16(4), 229.

[34] Kearns, V., Mac Intosh, A. C., Crawford, A., & Hatton, P. V. (2008). Silk-based biomaterials for tissue engineering. Topics in tissue engineering, 4, 1-19.

[35] Tamada, Y. (2005). New process to form a silk fibroin porous 3-D structure. Biomacromolecules, 6(6), 3100-3106.

[36] Hardy, J. G., Römer, L. M., & Scheibel, T. R. (2008). Polymeric materials based on silk proteins. Polymer, 49(20), 4309-4327.

[37] Grasset, L., Cordier, D., & Ville, A. (1977). Woven silk as a carrier for the immobilization of enzymes. Biotechnology and Bioengineering, 19(4), 611-618.

[38] Grasset, L. (1979). Silk: a natural protein for enzyme immobilization.

[39] Cordier, D., Couturier, R., Grasset, L., & Ville, A. (1982). Ribonuclease insolubilization using diazotized silk. Enzyme and Microbial Technology, 4(4), 249-255.

[40] Grasset, L., Cordier, D., Couturier, R., & Ville, A. (1983). Immobolization of alkaline phosphatase on silk using diazo, adsorption, glutaraldehyde, and azide method: Optimum pH and properties of the conjugates. Biotechnology and bioengineering, 25(5), 1423-1434.

[41] Dash R, Ghosh SK, Kaplan DL, Kundu SC. Purification and biochemical characterization of a 70 kDa sericin from tropical Tasar silkworm, *Antheraea mylitta*.Comp Biochem Physiol B: Biochem Mol Biol 2007;147:129-34.

[42] Yamada H, Tsubouchi K. Characterisation of silk proteins in the cocoon fibers of *Cricula trifenestrata*. Int J Wild Silkmoth Silk2001;6: 17-51.

[43] Dash R, Mukherjee S, Kundu SC. Isolation, purification and characterization of silk protein sericin from cocoon peduncles of tropical Tasar silkworm, *Antheraea mylitta*. Int J BiolMacromol2006;38: 255-8.

[44] Soong HK, Kenyon KR. Adverse reactions to virgin silk sutures incataract surgery. Ophthalmology 1984;91:479-83.

[45] Vicent MJ, Duncan R. Polymer conjugates: nanosized medicines for treating cancer. Trends Biotechnol 2006;24:39-47.

[46] Zhang YQ, Tao ML, Shen WD, Zhou YZ, Ding Y, Ma Y, et al. Immobilizationof l-asparaginase on the microparticles of the natural silk sericin protein and its characters. Biomaterials 2004;25:3751-9.

[47] Zhang YQ, Tao ML, Shen WD, Mao JP, Chen YH. Synthesis ofsilk sericin peptides–l-asparaginase (SS–ASNase) bioconjugates and their characterization. J Chem Technol Biotechnol 2006;81:136-45.

[48] Zhang YQ,MaY, Xia YY, ShenWD,MaoJP,Xue RY. Silk sericin–insulinbioconjugates: synthesis, characterization and biological activity. JControl Release 2006;115:307-15.

[49] Anghileri A, Lantto R, Kruus K, Arosio C, Freddi G. Tyrosinase catalyzed grafting of sericin peptides onto chitosan and production of protein–polysaccharide bioconjugates. J Biotechnol2007; 127:508-19.

[50] Yoshii F, Kume T, Makuuchi K, Sato F. Hydrogel composition containingsilk protein. Japan Patent 2000-169736A; 2000.

[51] Wang S, Goto Y, Ohkoshi Y, NaguraM. Structures and physical properties of poly(vinyl alcohol)/sericin blend hydrogel membranes. J Seric Sci Jpn 1998;67:295-302.

[52] Roy, P., Amdekar, S., Kumar, A., & Singh, V. (2011). Preliminary study of the antioxidant properties of flowers and roots of *Pyrostegiavenusta* (Ker Gawl) Miers. BMC Complementary and Alternative Medicine, 11(1), 69.

[53] Ananthi, P., &Kumari, B. D. (2013). GC–MS Determination of Bioactiv Components of Rorippaindica L. International Journal of Chem Tech Research, 5(4), 2027-2033.

[54] Dey, P., Ray, S., Sarkar, M. P., &Chaudhuri, T. K. (2015a). Chemical characterization and assessment of antioxidant potentiality of *Streptocaulonsylvestre* Wight, an endangered plant of sub-Himalayan plains of West Bengal and Sikkim. BMC complementary and alternative medicine, 15(1), 107.

[55] Chisholm, M. G., Wilson, M. A., &Gaskey, G. M. (2003). Characterization of aroma volatiles in key lime essential oils (*Citrus aurantifolia* Swingle). Flavour and Fragrance Journal, 18(2), 106-115.

[56] GomathiRajashyamala, L., &Elango, V. (2015). Identification of bioactive components and its biological activities of *Evolvulus alsinoides* linn.--A GC-MS study. IJCS, 3(1), 41-44.

[57] Premlata, S., Mourya, K. K., & Padma, K. (2015). Gas chromatography-mass spectrometric analyses of acetone extract of Marwar Dhaman grass for bioactive compounds. Plant Archives, 15(2), 1065-1074.

[58] Parthipan, B., Suky, M. G. T., & Mohan, V. R. (2015). GC-MS Analysis of Phytocomponents in *Pleiospermiumalatum* (Wall. ex Wight &Arn.) Swingle,(Rutaceae). Journal of Pharmacognosy and Phytochemistry, 4(1), 216-222.

[59] Zhang, H., Deng, L., Yang, M., Min, S., Yang, L., & Zhu, L. (2011). Enhancing Effect of Glycerol on the Tensile Properties of *Bombyx mori* Cocoon Sericin Films Enhancing Effect of Glycerol on the Tensile Properties of *Bombyx mori* Cocoon Sericin Films, (December). https://doi.org/10.3390/ijms12053170

[60] Iwara, I. A., Igile, G. O., Mboso, O. E., Mgbeje, B. I. A., &Ebong, P. E. (2017). Evaluation of phytochemical components from ethyl acetate fraction of *Vernoniacalvoana* using gas chromatography-mass spectrometry analysis and its antioxidants activities. African Journal of Pharmacy and Pharmacology, 11(42), 534-539.

[61] Lee, P. R., Yu, B., Curran, P., & Liu, S. Q. (2010). Kinetics of volatile organic compounds during papaya juice fermentation by three commercial wine yeasts. Nutrition & Food Science.

[62] Meechaona, R., Sengpracha, W., Banditpuritat, J., Kawaree, R., &Phutdhawong, W. (2007). Fatty acid content and antioxidant activity of Thai bananas. Maejo International Journal of science and technology, 1(2), 222-228.

[63] Aparna, V., Dileep, K. V., Mandal, P. K., Karthe, P., Sadasivan, C., &Haridas, M. (2012). Anti-inflammatory property of n-hexadecanoic acid: structural evidence and kinetic assessment. Chemical biology & drug design, 80(3), 434-439.

[64] Sen, T., &Samanta, S. K. (2014). Medicinal plants, human health and biodiversity: a broad review. In Biotechnological applications of biodiversity (pp. 59-110). Springer, Berlin, Heidelberg.

[65] Chong, Y. H., & Ng, T. K. (1991). Effects of palm oil on cardiovascular risk. Med J Malaysia, 46(1), 41-50.

[66] Ruhul, A. M., Kalam, M. A., Masjuki, H. H., Alabdulkarem, A., Atabani, A. E., Fattah, I. R., &Abedin, M. J. (2016). Production, characterization, engine performance and emission characteristics of *Croton megalocarpus* and *Ceibapentandra* complementary blends in a single-cylinder diesel engine. RSC advances, 6(29), 24584-24595.

[67] Selvamangai, G., &Bhaskar, A. (2012). GC–MS analysis of phyto components in the methanolic extract of *Eupatorium triplinerve*. Asian Pacific Journal of Tropical Biomedicine, 2(3), S1329-S1332.

[68] Teres, S., Barceló-Coblijn, G., Benet, M., Alvarez, R., Bressani, R., Halver, J. E., &Escriba, P. V. (2008). Oleic acid content is responsible for the reduction in blood pressure induced by olive oil. Proceedings of the National Academy of Sciences, 105(37), 13811-13816.

[69] Ismail, G. A., Gheda, S. F., Abo-shady, A. M., & Abdel-karim, O. H. (2020). In vitro potential activity of some seaweeds as antioxidants and inhibitors of diabetic enzymes. Food Science and Technology, 40(3), 681-691.

[70] Cicero, A. F., &Colletti, A. (2017). Food and plant bioactives for reducing cardiometabolic disease: How does the evidence stack up?. Trends in Food Science & Technology, 69, 192-202.

[71] Turro, N. J., Han, N., Lei, X. G., Fehlner, J. R., & Abrams, L. (1995). Mechanism of dichlorination of n-dodecane and chlorination of 1-chlorododecane adsorbed on ZSM-5 zeolite molecular sieves. A supramolecular structural interpretation. Journal of the American Chemical Society, 117(17), 4881-4893.

[72] Al-Wathnani, H., Ara, I., Tahmaz, R. R., Al-Dayel, T. H., &Bakir, M. A. (2012). Bioactivity of natural compounds isolated from cyanobacteria and green algae against human pathogenic bacteria and yeast. Journal of Medicinal Plants Research, 6(18), 3425-3433.

[73] Singariya, P., Mourya, K. K., & Kumar, P. (2015). Gas Chromatography-Mass Spectrometric Analyses of Acetone extract of Marwar Dhamangrass for bio-active compounds. Plant Archives. 2015a, 15(2), 1065-1074.

[74] Kang, W. Y., Ji, Z. Q., & Wang, J. M. (2009). Composition of the essential oil of *Adiantumflabellulatum*. Chemistry of natural compounds, 45(4), 575.

[75] Cardeal, Z. D. L., Gomes da Silva, M. D. R., & Marriott, P. J. (2006). Comprehensive two-dimensional gas chromatography/mass spectrometric analysis of pepper volatiles. Rapid Communications in Mass Spectrometry: An International Journal Devoted to the Rapid Dissemination of Up-to-the-Minute Research in Mass Spectrometry, 20(19), 2823-2836.

[76] Kaur, N., Shafiq, N., Negi, H., Pandey, A., Reddy, S., Kaur, H., & Malhotra, S. (2014). *Terminaliaarjuna* in chronic stable angina: Systematic review and meta-analysis. Cardiology research and practice, 2014.

[77] Smritikana Biswas, PranoyMitra. (2017). Comparative studies on the antimicrobial activity of *Terminalia arjuna* and *Aloe vera* against community associated drug resistant *Staphylococcus aureus* in pus from carbuncles of adults. Journal of Innovations in Pharmaceutical and Biological Sciences, 4 (4), 153-157

[78] Desai, D., &Chanda, S. (2014). Pharmacognostic Study and Physicochemical Analysis of Leaves of *Terminalia arjuna*. Pharmacognosy Journal, 6(6), 15-19.

[79] Ahamad, M. S. I., Neetha, K., & Vootla, S. K. (2020). Non-protein Chemical Compounds from Lepidopteran Insect Cocoons. In Natural Materials and Products from Insects: Chemistry and Applications (pp. 137-156). Springer, Cham

[80] Mandal, B. B., Ghosh, B., &Kundu, S. C. (2011). Non-mulberry silk sericin/ poly (vinyl alcohol) hydrogel matrices for potential biotechnological applications. International Journal of Biological Macromolecules, 49(2), 125-133. https://doi.org/10.1016/j.ijbiomac.2011.03.015

Chapter 5

Functional Morphology of Gustatory Organs in Caterpillars

Vonnie Denise Christine Shields

Abstract

The sense of taste plays a pivotal role in the behavior of insects. Caterpillars depend largely on taste cues from plants to detect and locate food sources. Taste stimuli can be either simple or complex as multimolecular mixtures. The insect faces the task of deciphering the nature of these tastants and must then make appropriate feeding choices. Typically, caterpillar larvae possess four types of bilateral gustatory sensilla on their mouthparts. The lateral and medial styloconic sensilla are thought to be the primary organs involved in feeding. These sensilla are in continuous contact with plant sap during feeding and can detect different phytochemicals present in the plant. The gustatory sensory input is encoded as patterns of nerve impulses by gustatory receptor cells housed in these sensilla. Therefore, these gustatory receptor cells form the first layer of a decision-making process that ultimately determines whether food is accepted or rejected by the insect. Caterpillars, such as gypsy moth larvae (*Lymantria dispar*) (L.) (Lepidoptera: Lymantriidae) are major forest pests in most of the United States. These larvae are highly polyphagous feeders and defoliate a variety of tree species, including forest, shade, fruit, and ornamentals. This chapter discusses morphological, feeding behavioral, and electrophysiological aspects of gustatory sensilla with respect to gypsy moth caterpillars.

Keywords: gustation, taste, ultrastructure, insect plant interactions, feeding behavior, electrophysiology

1. Introduction

Gustation is crucial for the survival and nutrition of animals. It is critical in determining the palatability of foods and in providing early warning signs of spoilage. This chapter promotes a better understanding of how natural taste (gustatory) stimuli are recognized, coded, and processed by receptor cells housed in gustatory sensory organs (sensilla) using an insect model, gypsy moth caterpillars, *Lymantria dispar* (L.). These sensilla are cuticular structures which house gustatory receptor cells in them. These receptor cells constitute a sensory filter for environmental taste signals. In insects, these receptors transfer information directly to higher processing taste centers in the brain and form the first layer of a decision-making process which determines if food should be accepted or rejected. Typically, the insect faces the task of deciphering individual tastants in a complex multimolecular mixture to make appropriate feeding choices. In order to respond to stimuli in different behavioral or ecological contexts and to discriminate between meaningful taste stimuli, caterpillars (larvae) have evolved several different types of gustatory sensilla.

Food plant recognition is predominantly governed by the activity of two pairs of sensilla located on the mouthparts, namely the lateral and medial styloconic sensilla [1–3]. When the larva feeds, these gustatory sensilla are in continuous contact with the plant sap and can detect different chemicals (i.e., phytochemicals, secondary plant compounds, allelochemicals) present in the plant. Larval gustatory sensilla provide an excellent system to address questions about the taste system, since: i) these sensilla form a relatively simple sensory system with a limited number of sensory cells that mediate gustatory mechanisms; ii) these sensilla are readily accessible for experimental manipulation, and iii) the receptor cells within these sensilla are individually identifiable and exhibit typically robust and reproducible electrophysiological responses [1].

2. Chemosensory systems and sensillum types

Adult insects possess several different types of sensilla that monitor the environment for cues associated with finding food, oviposition sites, conspecific mates, suitable temperature and humidity levels, and seeking protection and orientation. These sensory organs enable them to detect stimuli associated with taste, smell, touch, sound, vision, proprioception, and geo-, thermo-, and hygroreception. In contrast, the sensory requirements of larvae, such as those found in the order Lepidoptera, are more limited. For example, they rely strongly on gustatory, tactile, and possibly short-range olfactory cues for host-plant selection [4]. Lepidopterous insects use various physical and chemical characteristics to locate plants. Although the visual sense aids a caterpillar to reach a plant, this sense is not finely enough developed to play a role in food plant recognition. The chemical senses, which are well developed in insects, not only guide monophagous insects (feed on only one or a few closely related plant species) to its food, but also helps polyphagous insects (feed on many plants belonging to different plant families) to discriminate various plant species. Chemoreceptors are located on the antennae and mouthparts (**Figure 1**). In total, lepidopterous larvae have five types of bilateral chemosensilla found on the head: a pair of antennae (each innervated by 16 neurons), two pairs of lateral and medial styloconic sensilla located on the galea (each pair innervated by eight neurons), a pair of maxillary palps (each with eight sensilla on their distal surface, and each innervated by 14–19 olfactory and gustatory neurons), and a pair of epipharyngeal organs (each innervated by three gustatory neurons) [1].

Three main categories of insect sensilla exist: (1) AP (aporous) or NP (no-pore) sensilla, which are either mechanosensitive or hygro- and thermosensitive; (2) UP (uniporous) or TP (terminal-pore) sensilla containing gustatory neurons, alone, or with a mechanosensitive cell, and (3) MP (multiporous) or WP (wall pore) sensilla (single-walled (SW) sensilla and double-walled (DW) sensilla). Often, multiporous sensilla are olfactory and wall pore are olfactory and/or thermohygrosensitive [6, 7]. The lateral and medial styloconic sensilla are uniporous or terminal pore sensilla.

3. Lateral and medial styloconic sensilla

The sense of taste in insects is referred to as contact chemoreception. Contact chemosensilla are analogous to the taste buds located on the tongue in the oral cavity of vertebrates. In lepidopterous larvae, gustatory sensilla are located on the mouthparts, specifically the maxillae and epipharynx [6–9]. Each maxilla is comprised of a maxillary palp and galea. Each galea bears two elongated protuberances, namely the lateral and medial styloconic sensilla (**Figures 1** and **2**). These sensilla are located

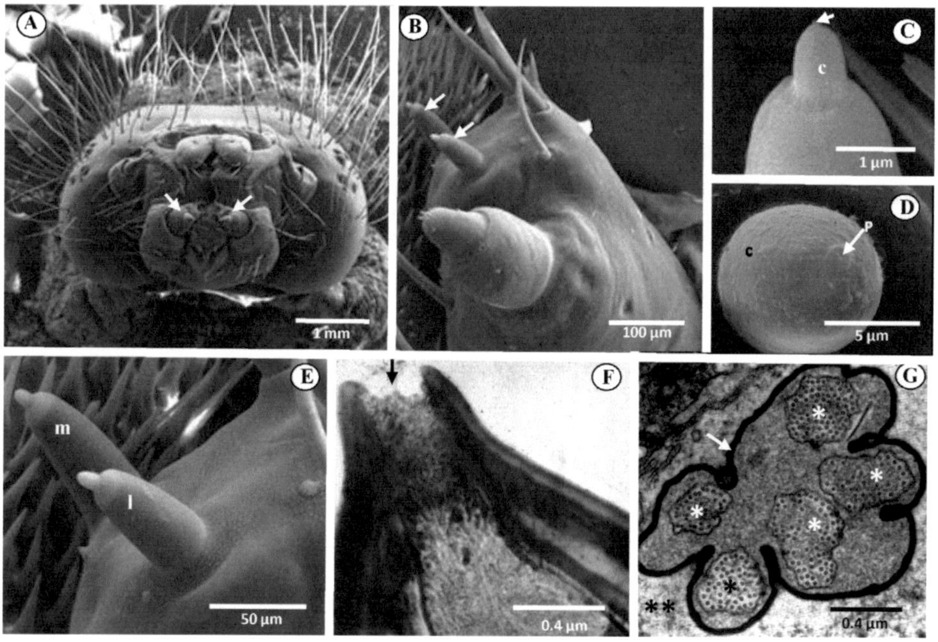

Figure 1.
A-E, scanning electron micrographs and F, G, transmission electron micrographs of Lymantria dispar (L.) *fifth instar larvae. The specimens shown A-E, critical point dried. A) Frontal view, whole head. The arrows point to the galeae, components of the maxillae. Bar = 1 mm. B) Superior-dorsal view of the tip of a left galea showing lateral and medial styloconic sensilla (arrows). Bar = 100 µm. C) Side view of a medial styloconic sensillum showing a higher magnification of a cone (c) inserting into the style (cylindrical projection beneath the cone). The arrow denotes the location of a terminal pore. Bar = 1 µm. D) Higher magnification view of the apical view of a cone (c) showing the terminal pore (p with arrow) from a lateral styloconic sensillum. Bar = 5 µm. E) Higher magnification view of figure B showing the lateral (l) and medial (m) styloconic sensilla. Bar = 50 µm. F) Longitudinal section of a lateral styloconic sensillum showing the tip of the pore (arrow), which contains an apparent plug of fenestrated fibrils. Bar = 0.4 µm. G) Cross section taken near the base of the cone, proximal to where it inserts into a long cylindrical projection (style), and proximal to the site of the tubular body within the mechanosensory dendrite. The five distal dendrites (asterisks) (four chemosensory and one mechanosensory) within the dendritic channel and are surrounded by the conspicuous electron-dense dendritic sheath (arrow) and sensillar sinus (double asterisks). Bar = 0.4 µm. From [5].*

near the mouth opening of the caterpillar. During feeding, these sensilla come into continuous contact with the plant sap before it enters the mouth or buccal cavity and can detect different chemicals present in the plant sap (i.e., allelochemicals) [10]. In lepidopterous larvae, food plant recognition is thought to be primarily mediated by the input from each bilateral pair of styloconic sensilla [1–3, 10–14]. Therefore, they are considered the primary sensory organs involved in feeding [15–18].

Ablation experiments have shown that removal of the styloconic sensilla of the tobacco hornworm, *Manduca sexta,* resulted in widening of its host range [19]. Ablation of the styloconic sensilla in this species permitted it to feed on previously unacceptable plants, thus broadening the range of tolerable plant species of this insect [3]. It was concluded that rejection of plants could be mediated by either the medial or lateral styloconic sensillum and that both sensilla were involved in the rejection behavior to different substances [20, 21]. Thus, these results support the notion that receptor cells present in each styloconic sensillum are involved in selectively mediating the blocking of feeding behavior. Brightfield light microscopic studies, as well as transmission electron microscopy, have revealed that these sensilla each bear a single permeable apical pore (uniporous, UP, or terminal pore, TP sensilla) and are typically innervated by five bipolar neurons, four of which function as putative gustatory receptors and one, as a putative

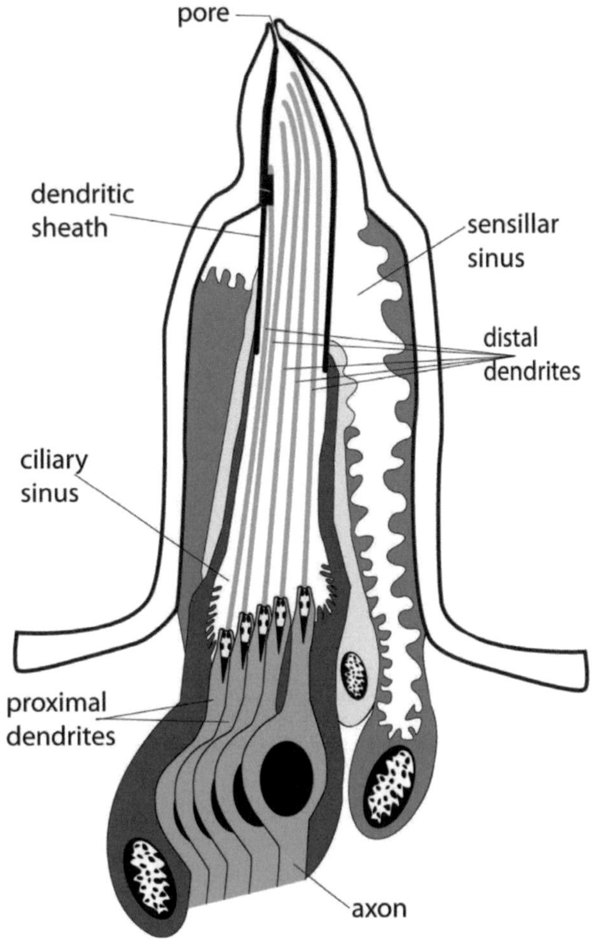

Figure 2.
Diagrammatic reconstruction of a uniporous styloconic sensillum of the gypsy moth, Lymantria dispar, *shown in longitudinal section. Five bipolar neurons are present (four gustatory, one mechanosensory). The dendritic sheath completely separates the dendrites within the dendritic channel from the large sensillar sinus. The sheath ends distal to the small ciliary sinus. In the ciliary region, the distal dendritic segments insert into the proximal dendritic segments. Modified from [9].*

mechanoreceptor [1, 6, 8, 9, 22] (**Figure 2**). The styloconic sensillum is so named, since it appears as a small cone, peg, or knob-like structure that is inserted into a cylindrical projection (style) of insensitive cuticle [23] (**Figures 1** and **2**) and is, therefore, classified as a uniporous (UP) sensillum [6, 7].

4. Structure of uniporous styloconic sensilla

A uniporous styloconic sensillum can take the form of a short to medium-long peg or cone that is inserted into a fibrous cuticular socket of the style, which allows it to flex in this articular region (**Figures 1** and **2**). A schematic reconstruction of a styloconic sensillum is shown in **Figure 2**. The sensillum bears a single permeable apical pore located at the tip and is typically innervated by five bipolar neurons, four of which function as putative gustatory receptors and one, as a putative mechanoreceptor [1, 8, 9, 24]. The pore, about 10–200 nm in diameter, contains typically pore tubules or plugs of fenestrated fibrils [8, 9, 25] allowing chemical communication to occur between the receptor cells and the external environment.

The pore fibrils may also confer selectivity to the conduction mechanism and specificity of response to the sensillum [6, 25]. Four putative gustatory neurons extend within a dendritic channel inside the sensillum from the pore. A dendritic sheath encloses the dendrites. This sheath extends from near the tip of the sensillum to approximately the level of the ciliary sinus. This sinus bathes the dendrites. This sheath completely separates the dendrites from a large sensillar sinus. The dendritic sheath, possibly perforated by pores in some regions, could enable the sensillar sinus to act as reservoir of ions and resting potentials, as has been shown for taste sensilla of adult flies [7, 25, 26]. The fifth putative unbranched mechanosensory dendrite begins near the base of the cone and lies closely apposed to the dendritic sheath and cuticular wall of the cone. The apical termination of this dendrite bears an accumulation of microtubules. These microtubules lie parallel to one another within an electron-dense matrix (tubular body) and is thought to be the site of sensory transduction of mechanical stimuli [27]. The dendrites constrict abruptly midway along their lengths in the ciliary region. This point distinguishes the distal dendritic (ciliary) segments from the proximal dendritic segments. The proximal dendrites continue proximally and form cell bodies. From this point, axons from the lateral and medial styloconic sensilla merge and form the lateral and medial branches of the galeal nerve and project directly without synapsing into the subesophageal ganglion (SOG) [28, 29]. The SOG is thought to serve as the first order relay station in the central nervous system. The SOG also exerts motor control over the mouthparts that are directly involved in the feeding process [30–32]. Much of the central processing of various types of input (including gustatory cells) takes place in the SOG, however since inputs from other parts of the central nervous system (e.g., frontal ganglion, olfactory lobes) also contribute to feeding behavior (i.e., host-plant recognition), it is unclear if the "feeding center" is wholly situated in the SOG [33].

5. Feeding behavior

All insects are selective to some extent in their food choice, feeding on (a) one or a few closely related plant species (monophagy), (b) a larger number of hosts usually confined within a certain plant family (oligophagy) or (c) many plants representing a wide taxonomic range (polyphagy). Insects never feed on all plant groups, however [34]. The main function of contact chemoreceptors on the mouthparts of insects is the selection of food. When an insect bites into a plant, some contact chemoreceptors become exposed to the plant sap and function similarly to taste receptors in vertebrates by detecting the compounds in solution [11–14]. However, some mouthpart sensilla, such as would be found in lepidopterous larvae (e.g., found on the maxillary and labial palps), often contact the food before the insect bites. The receptors within these sensilla are sensitive to compounds on the dry surface of a leaf when these sensilla are brought into brief contact with the plant surface. This palpatory behavior serves to: (1) allow the insect to receive a more sustained flow of information from the receptors than would be possible if contact were maintained, since the receptors would become adapted and (2) allow the insect to sample a greater leaf surface than if the sensilla would have remained stationary [35]. The information obtained by palpation, therefore alerts the insect to avoid the intake of noxious compounds and to make feeding decisions more rapidly.

Food selection behavior should be compared to a "key-lock" system where the key represents a receptor activity profile [36]. Only when this profile sufficiently corresponds with an innate standard in a pattern recognition area in the central nervous system is a particular behavioral response triggered. When the incoming sensory information differs too much from the desired pattern, the food is rejected. The central nervous system (or lock), consisting of the SOG and other brain regions, is tuned to recognize sensory patterns. Those patterns recognized as acceptable will release feeding behavior, while others will result in food rejection. The final decision is thought to be made in the SOG. In the case of a specialist feeder, the incoming sensory pattern would have to match more closely a certain norm set by the central nervous system to trigger feeding activity, whereas in a generalist, many different receptor activity profiles can evoke a feeding response. In order to understand feeding behavior, it is necessary (a) to examine which allelochemicals elicit an acceptance or rejection response and (b) to determine the function and number of taste receptor cells within the styloconic sensilla that are involved in mediating acceptance or rejection of food plants, and (c) to describe how the receptor cells housed in these sensilla encode this taste information and transmit it to the central nervous system to prevent (deter) or elicit feeding behavior.

6. Phytochemicals and hostplant preferences

Phytochemicals include primary and secondary plant metabolites. Secondary plant substances (i.e., allelochemicals) are not universally found in higher plants, but are restricted to certain plant taxa (or occur in those taxa at much higher concentrations than in others) and are of no nutritional significance to insects [37, 38]. Plants produce a wide range of secondary metabolites that act as defense compounds from herbivores, as well as microorganisms. In addition, they can serve as attractants for pollinators. Still others share structural similarities to neurotransmitters [39]. Many secondary metabolites may be cytotoxic as they interfere with biomembranes, cytoskeletal proteins or DNA, and can induce apoptosis [40]. Food specificity can be based solely on the presence or absence of secondary metabolites. In certain plant taxa, these compounds can serve also as "sign" stimuli for some specialized insect species allowing them to unambiguously identify their hostplant, as well as act as effective defensive barriers against non-adapted species [34]. Deterrents (secondary plant substances that inhibit feeding) play important roles in host-plant interactions. It has been postulated that hostplant selection or hostplant acceptability is due to the lack of compounds present that inhibit feeding, whereas rejection of non-hostplants is due to the presence of feeding inhibitors or deterrents. The lack of compounds that inhibit feeding and rejection of non-hostplants is due to the presence of feeding inhibitors or deterrents [41]. The term "allelochemic" was coined and defined as a "non-nutritional chemical" that is produced by an individual of one species (plant) that affects the growth, health, behavior, or population of another species (insects) [37]. Commonly, a plant may produce more than a single allelochemical, which are stored at important sites in the plant [42].

Gypsy moth larvae display a wide host-plant preference [43]. They are highly polyphagous feeders (feed on many plants belonging to different plant families) and defoliate many tree species, including forest, shade, fruit, and ornamentals [44]. For polyphagous ("generalist") insect species, such as the gypsy moth, there may be a balance that exists between phagostimulants and deterrents which determine the extent to which a plant will be eaten or rejected [5, 45–48]. While phagostimulation is necessary to drive feeding, it is not likely to influence hostplant selection [35]. Therefore, hostplant selection is likely defined by the presence of deterrent compounds in non-hosts. Polyphagous insects are deterred from feeding on plants that store noxious metabolites and usually select those with less active ones [49]. Alternatively, they may also avoid intoxication by changing hostplants rapidly and have evolved detoxification and rapid excretion mechanisms for certain allelochemicals [49, 50]. In contrast, for many oligophagous (feed on several plant species, belonging to the same plant family) and all monophagous (feed on only one or a few closely related plant species) ("specialist") insect species, feeding appears to be driven by the presence of chemicals that act as "sign stimuli." That allow the insect to unambiguously identify their hostplant and stimulate feeding, as well as act as effective defensive barriers against non-adapted species and identify the presence of deterrent compounds in non-host plants [35, 49]. These "sign stimuli" may have been originally noxious but can be tolerated (detoxified) and/or sequestered for the insect's defense against predators or show a relative lack of deterrent effects in the hostplant [35, 49]. Gypsy moth larvae are "generalist" feeders and capable of destroying entire forests during outbreak years. Relatively few studies have documented which allelochemicals are relevant in eliciting acceptance or rejection feeding responses in this generalist herbivore (e.g., [43, 46–48, 51–54]. There is only one study to date that has described the detailed ultrastructural morphology and sensory physiology of chemoreceptors housed within the maxillary galeal styloconic sensilla, thought to be the primary organs involved in feeding [9]. Consequently, our knowledge of the basic mechanisms of chemoreception of gypsy moth larvae lags that of other lepidopterous larvae, such as *Manduca sexta*, *Pieris brassicae*, and *Bombyx mori* (reviewed in [35, 55]. Two-choice feeding behavioral bioassays using *L. dispar* caterpillars revealed that plants containing alkaloids, one of the largest chemically heterogenous groups of allelochemicals, occurring in 20–30% of higher plants, were unfavored by gypsy moth larvae (**Figures 3** and **4**) [43, 46, 47].

Figure 3.
Experimental set-up for two-choice feeding behavioral bioassay showing the arrangement of control (A) leaf disks and those treated with an alkaloid (B). The disks were punched out of red oak, Quercus rubra (L.), *a plant species highly favored by* L. dispar *larvae and arranged in an alternating circular fashion (ABABAB) (technique modified after [41]. Metal pins were pushed through the center of each disk into dental wax to ensure that the disks stood ca. 5 mm above the wax surface. The test compounds were dissolved in appropriate solvents and applied so that the chemical amounted to 1% of the dry weight of the disk. Experiments were run until 50% of total area of either control or test disks were consumed. Leaf disks were oven-dried following each experiment for 48 h and then weighed. Values were reported as percent relative mean consumption of control consumption.*

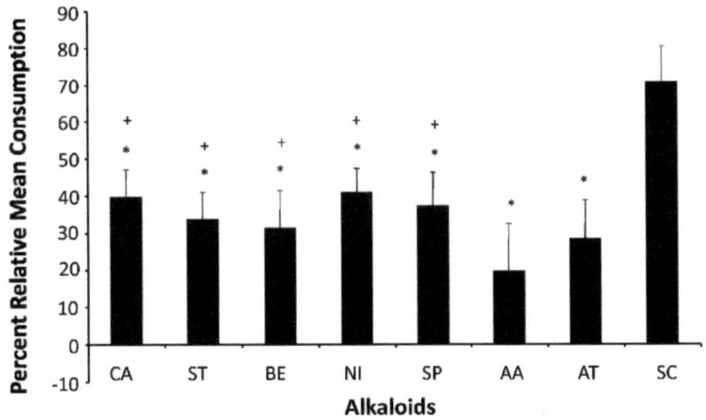

Figure 4.
Two-choice feeding bioassay showing the results of percent relative mean consumption of eight selected alkaloids when applied to red oak leaf disks by fifth instar L. dispar *larvae. Consumption was normalized with respect to control disks (100%). Bars represent the alkaloids tested. AA aristolochic acid, AT atropine, BE berberine, CA caffeine, NI nicotine, SC scopolamine, SP sparteine, ST strychnine. Results are derived from 23, 25, 15, 34, 30, 34, 21, and 15 larvae (number of replicates). Asterisks indicate alkaloids that significantly deterred feeding ($P < 0.05$). Plus symbols indicate alkaloids that were significantly less deterrent on red oak leaves compared with glass fiber disks, i.e., red oak leaves reduce alkaloid deterrent effects. Error bars represent S.E. from [46].*

7. Taste receptor cell classification and peripheral gustatory coding

Insects, like other animals, can taste major nutrients essential for their development, survival, and reproduction, including sugars and inorganic salts. Lepidoptera typically use separate cells that are sensitive to a wide range of chemicals to mediate information about the presence of chemicals, including sugars ("sugar best cell"), inositol ("inositol best cell"), salts ("salt best cell"), and deterrents ("deterrent cell") [38]. In humans, the latter compounds would taste "bitter" [56]. This classification does not necessarily imply that these cells respond only to these groups of chemicals but are more sensitive to them and are likely to be activated by them.

Sensory inputs from food elicit behavioral responses in insects. There is no direct experimental evidence, however, how inputs from taste receptors are integrated in the central nervous system. There appears, however, to be a direct relationship between the amount eaten and the activity of taste receptor cells to different concentrations of a stimulant. Conversely, as the activity of the deterrent cell increases with concentration of the deterrent, the amount eaten declines [55]. It is presumed that these inputs are brought together in the central nervous system in an additive manner and have positive effects. Deterrents, on the other hand, have negative effects on feeding. Insect gustatory receptors transduce the quality and quantity of the complex plant chemistry into a neural code of action potentials. Complex stimuli resulting from e.g., plant saps often evoke spike trains in several receptor cells innervating one or more sensilla. Typically, each cell type (e.g., sugar best cell versus deterrent cells) can be distinguished based on its spike template and temporal firing pattern (**Figures 5–7**) [35, 58]. The frequency and temporal distribution of action potentials in a spike train contains information about the stimulus. The axons project to and converge in the first relay station, the SOG, without intermittent synapses. Unraveling the sensory code occurs by analyzing "input–output" relationships [58, 59]. This can be achieved by stimulating specific sensilla and quantifying electrophysiological recordings of the trains of action potentials (input), as well as quantifying the behavior (output) based on how much food is consumed [35]. Coding is inferred by making correlations between input and output.

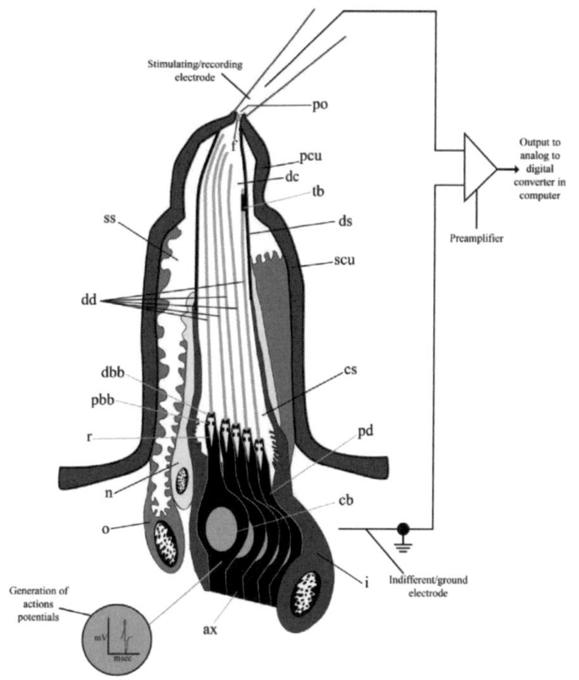

Figure 5.
Diagrammatic reconstruction of a uniporous styloconic sensillum in longitudinal section. This illustration shows, in addition, the electrophysiological tip recording method [57] used to record the excitatory responses from individual taste cells found within a styloconic sensillum. All five sensory cells are shown in this reconstruction. The stimulating or recording electrode contains the taste stimulus dissolved in an electrolyte solution (e.g., 0.1 M KCl dissolved in deionized water). This electrode is placed over the tip and terminal pore of a styloconic sensillum. The solution then diffuses through the terminal pore of the sensillum and taste compounds bind to dendritic taste receptors which transduce the quality and quantity of the taste stimulus into a neural code of action potentials. The other electrode, the indifferent or ground electrode, also contains a similar electrolyte solution and is positioned to make contact with the internal environment of the insect (e.g., body). Both electrodes contain, in addition, a silver wire. The excitatory responses are then recorded, amplified, digitized, and analyzed using a computer software program. Ax, axon; cb, cell body; cs, ciliary sinus; dbb, distal basal body of proximal dendritic segment; dc, dendritic channel; dd, distal dendritic segment; ds, dendritic sheath; f, fibrils; i, inner sheath cell; n, intermediate sheath cell; o, outer sheath cell; pcu, peg cuticle; pd., proximal dendritic segment; po, terminal pore; pbb, proximal basal body of proximal dendritic segment; r, rootlets; scu, style cuticle; ss, sensillar sinus; tb, tubular body. From [5].

To better comprehend the neural communication between chemosensory organs and the central nervous system resulting in acceptance or rejection behavior, three theories exist to best describe the sensory responses: (1) labeled line, (2) across-fiber patterning, and (3) temporal patterning. The first theory proposes that the more important a single compound is in controlling or modifying behavior, the more likely its detection will be coded by a single cell [60]. This "labeled line" (i.e., line or axon along which information is transferred to the brain) to the central nervous system would only carry information from cells with a narrow and well-defined sensitivity spectrum of a specific chemical (or family of chemicals) and would be directly linked to a specific behavioral response [55]. The second theory suggests that the nervous system bases its decision for behavioral output by evaluating the responses from many individual sensory cells with different but overlapping response spectra. The central nervous system extracts meaningful information by reading and processing simultaneous inputs across all afferent sensory fibers (axons) (across-fiber patterning) [61]. This is also known to occur in vertebrates [17]. The third theory implies that temporal patterning may be superimposed on across-fiber patterning suggesting that the ratios of firing across different cells changes with time and can modify a particular

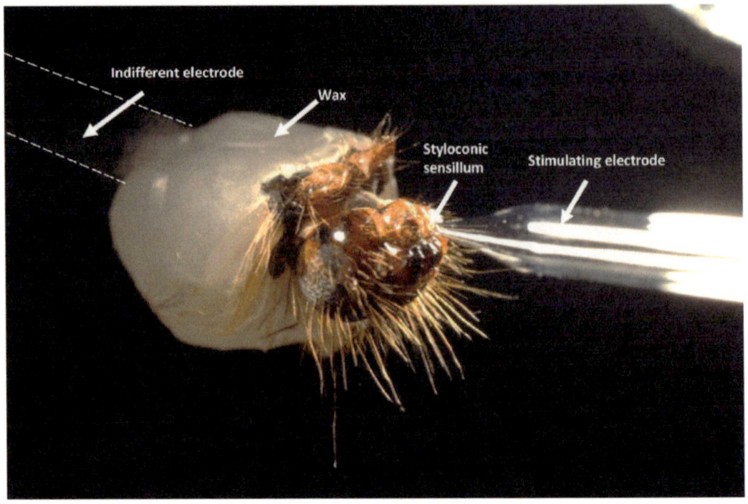

Figure 6.
Photograph of the electrophysiological tip-recording technique as explained in more detail in ***Figure** 2. The stimulating electrode, containing the taste stimulus and dissolved in an electrolyte solution, is placed over the tip of a styloconic sensillum. The indifferent electrode, containing a similar electrolyte solution, is inserted into the body of the insect. A minimal amount of melted wax is used to secure the preparation. From [5].*

message [62]. Most importantly, it should be noted that all three theories (code types) are not mutually exclusive and can be combined into one model [63].

Sensory codes mediating acceptance can: (i) stimulate specific sugar cells coding for an acceptance profile; (ii) stimulate broad spectrum sugar cells that the CNS recognizes as an acceptance profile [62, 64] and (iii) inhibit specific phagodeterrent receptors; this contributes to the neural coding of acceptance [65]. Feeding deterrents may alter sensory input by: (i) stimulating specific deterrent receptors; (ii) stimulating broad spectrum receptors; (iii) stimulating some cells and inhibiting others, thereby changing complex and subtle codes; (iv) inhibiting specific phagostimulant receptors; this contributes to the neural coding of deterrence, and (v) evoking highly unnatural impulse patterns, often at high frequency [65]. The ability of a deterrent neuron to respond to a wide range of chemicals is due to it having a diverse range of receptor sites, each with its own structure–function specificity, or due to the active chemicals having common features making them able to interact with a single receptor site [66]. Deterrent cells possess a number of unique characteristics: (i) they generally adapt more slowly than cells responding to phagostimulatory compounds; (ii) the tonic activity of the deterrent receptor stabilizes at a higher level than in other cell types; (iii) there may be a relatively long latency period prior to the tonic response; (iv) there may be a slow increase in spike frequency following stimulus application, and (v) there may be an increase in spike amplitude with stimulus concentration [62, 67]. Differential adaptation rates are, thus, useful in explaining how a sensory code changes with time and how deterrent receptor activity gradually becomes more pronounced when the sensory message is sent to the brain [62]. Food, which at the beginning of a meal may be acceptable, soon becomes unacceptable because of the more prominent share of the deterrent in the total sensory impression. Using *Pieris brassicae* as a model, it was determined, that impulses from receptor cells that convey deterrent information are given a greater weight by the CNS [55]. Therefore, one impulse from a deterrent-sensitive neuron may neutralize 2.5 impulses from sugar sensitive cells. Furthermore, cells signaling the presence of allelochemicals usually respond to about 1000 times lower concentrations than the receptors measuring the quantity of nutrients.

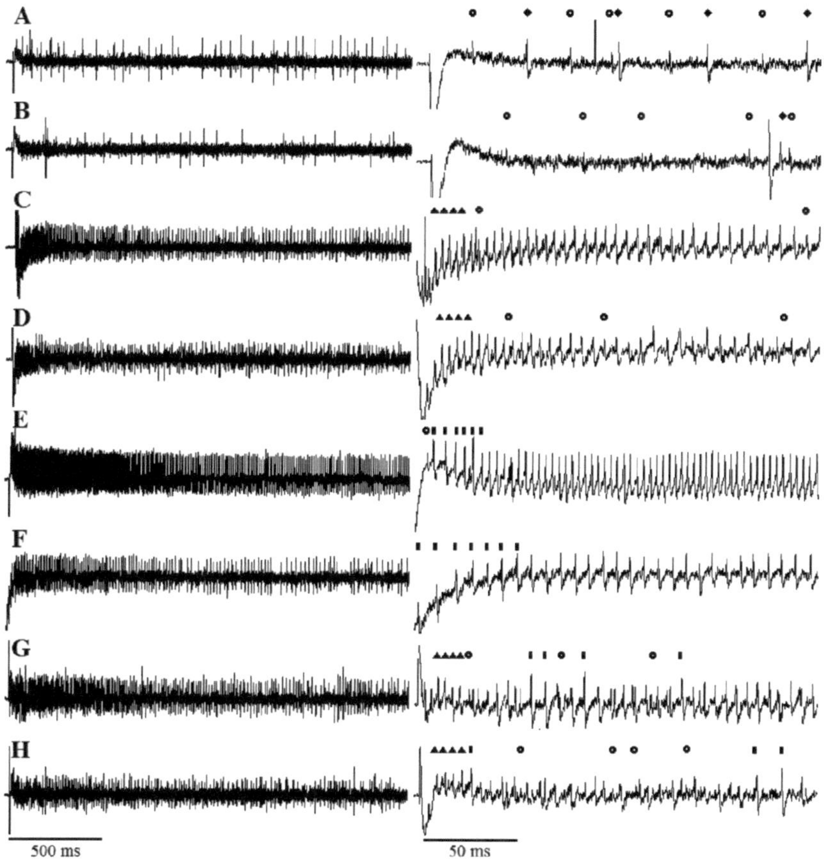

Figure 7.
Representative neurophysiological responses from medial styloconic sensilla of Lymantria dispar *in response to single-component taste stimuli, as well as binary mixtures. The recordings on the left side show only the first 2 s of stimulation. The recordings on the right side are identical to those on the left side but show only the first 200 ms of stimulation. All recordings were made using the same animal preparation. Response of a medial styloconic sensillum to A) 30 mM potassium chloride (salt), B) 100 mM sucrose (sugar) in 30 mM potassium chloride, C) 100 mM inositol (sugar alcohol) in 30 mM potassium chloride, D) the binary mixture of both sucrose and inositol in 30 mM potassium chloride, E) 1 mM strychnine (alkaloid) in 30 mM potassium chloride, F) the binary mixture of 1 mM strychine and 100 mM sucrose in 30 mM potassium chloride, G) the binary mixture of 1 mM strychnine and 100 mM inositol in 30 mM potassium chloride, and H) the mixture of 1 mM strychnine, 100 mM sucrose, and 100 mM inositol in 30 mM potassium chloride. The open circles and filled diamonds represent the firing of the two salt-sensitive cells; the filled triangles, the inositol-sensitive cell, and the filled rectangles, the deterrent-sensitive cell. In A), two taste cells (a smaller and a taller amplitude cell) fired independently of one another in response to potassium chloride. The appearance of a possible third cell in the recording on the right (spike not denoted by an open circle or filled diamond) is the result of both salt-sensitive cells firing at the same time. In B), a sucrose-sensitive cell was absent in medial styloconic sensilla, so only two cells fired in response to potassium chloride, like the response in a. the appearance of a possible third cell in the recording on the right (spike not denoted by an open circle or filled diamond) is the result of both salt-sensitive cells firing at the same time. In C), an inositol-sensitive cell fired in response to inositol. In D), the binary mixture of sucrose and inositol elicited the response of only the inositol-sensitive cell to inositol. The firing rate and amplitude of the inositol-sensitive cell was decreased with the addition of sucrose, implying a mixture-interaction effect. In E) a deterrent-sensitive cell fired large amplitude spikes in response to strychnine. In F), the binary mixture of strychnine and sucrose elicited the response of only the deterrent-sensitive cell in response to strychnine. The firing rate and amplitude of the deterrent-sensitive cell was decreased with the addition of sucrose, implying a mixture-interaction effect and that sucrose ameliorated the deterrent effect of strychnine. In G), the binary mixture of strychnine of inositol elicited the responses of two cells: The deterrent-sensitive cell and the inositol-sensitive cell. The firing rate and amplitude of the deterrent-sensitive cell was decreased with the addition of inositol, implying a mixture-interaction effect and that inositol ameliorated the deterrent effect of strychnine. In H) the mixture of strychnine, sucrose, and inositol elicited the responses of two cells: The deterrent-sensitive cell and the inositol-sensitive cell. The firing rate and amplitude of both cells was decreased with the addition of sucrose, implying a mixture-interaction effect. The addition of both sucrose and inositol ameliorated the deterrent effect of strychnine. From [5].*

8. Conclusions

Insects make ideal models for addressing the mechanisms that govern feeding behavior. As mentioned previously, the gustatory sensilla of lepidopterous caterpillars provide an excellent system to address questions about the taste system. These sensilla form a i) relatively simple sensory system with a small number of sensory cells that mediate gustatory mechanisms, ii) the location of these sensilla provides relatively easy access for experimental manipulation, and iii) the receptor cells within these sensilla are individually identifiable and exhibit typically reproducible electrophysiological responses. The anatomical organization and the molecular signaling pathways in taste are distinctly different between vertebrates and invertebrates (i.e., insects). Nevertheless, in both animal groups, the coding of taste quality has revealed surprising similarities, such that each of the taste qualities is mediated by a labeled line [68]. This means that a particular population of taste receptor cells is set apart and is responsible for encoding a specific taste quality.

At the molecular level, recent research with *Bombyx mori* has revealed that three putative bitter insect gustatory receptors (GRs) (BMGr16, BmGr18, and BmGr53) respond widely to structurally different and partially overlapping deterrents, suggesting that these bitter GRs are feeding deterrent receptors and play important roles in hostplant recognition [69]. Interestingly, feeding preference studies with *B. mori* have shown that the GR66 gene, encoding a putative GR, is responsible for the feeding preference on mulberry of this monophagous insect. With the aid of clustered regularly interspaced short palindromic repeats (CRISPR/CRISPR-associated protein-9-nuclease (Cas9) system, a mutation was introduced in the GR66 locus. As a result of this genetic mutation, *B. mori* larvae broadened their feeding activity. The larvae fed on several plant species not normally in their diet, leading to the discovery of the first genetic and phenotypic evidence that a single bitter GR can affect this insect's feeding preference [70]. The recent progress in functional genomics and molecular advances on bitter GRs of *B. mori*, points to new directions and strategies for controlling pest damage. Furthermore, it broadens our understanding about insect-plant interactions and yields new information about how insects perceive and process taste information.

Acknowledgements

This work was supported by NIH grants 1R15DC007609-01 and 3R15DC007609-01S1 to V.D.C.S. and grants from Towson University (Fisher College of Science and Mathematics Undergraduate research grants, Towson University Office of Undergraduate research grants, and NIH grant DC-02751). The author gratefully acknowledges J. Klupt, R. Kuta, T. Maugel, T.L. Martin, N.S. Arnold, B.P. Broomell, J.O.B. Salako, E.J. Rodgers, D. Williams, K.P. Smith, I.M. Gordon, T.E. Shaw, D. Waranch, B. K. Mitchell, B. Bennett, and USDA-APHIS (Falmouth, Massachusetts).

Author details

Vonnie Denise Christine Shields
Department of Biological Sciences, Towson University, Towson, MD, USA

*Address all correspondence to: vshields@towson.edu

References

[1] Schoonhoven, LM, Dethier, VG. Sensory aspects of host-plant discrimination by lepidopterous larvae. Arch. Néerl. Zool. 1966;16:497-530. DOI:10.1163/036551666X00057

[2] de Boer, G, Dethier, VG, Schoonhoven, LM. Chemoreceptors in the preoral cavity of the tobacco hornworm, *Manduca sexta*, and their possible function in feeding behaviour. Entomol. Exp. Appl. 1977;21: 287-298. DOI:10.1111/j.1570-7458.1977.tb02683.x

[3] de Boer, G and Hanson, FE. Differentiation of roles of chemosensory organs in food discrimination among hosts and non-host plants by larvae of the tobacco hornworm, *Manduca sexta*. Physiol. Entomol. 1987;12:387-398. DOI:10.1111/j.1365-3032.1987.tb00765.x

[4] Städler, E, Hanson, FE. Olfactory capabilities of the "gustatory" chemoreceptors of the tobacco hornworm larvae. J. Comp. Physiol. 1975;104:97-102. DOI:10.1007/BF01379454

[5] Shields, VDC, Martin, TL. The Structure and Function of Taste Organs in Caterpillars. In: Lynch, EJ and Petrov, AP, editors. The Sense of Taste, Nova Science Publishers, Inc. Hauppauge, NY. Chapter 11, pp. 147-166. 2012.

[6] Zacharuk, RY. Antennae and sensilla. In: Kerkut ,G.A. and Gilbert, L.I., editors. Comprehensive insect physiology, biochemistry and pharmacology. Vol. 6. Pergamon Press, Oxford, 1985. P. 1-69

[7] Zacharuk, RY, Shields, VD. Sensilla of immature insects. Annu. Rev. Entomol. 1991;36:331-354. DOI: 10.1146/annurev.en.36.010191.001555

[8] Shields, Vonnie D.C. 1994. Ultrastructure of the uniporous sensilla on the galea of larval *Mamestra configurata* (Walker) (Lepidoptera: Noctuidae). *Can. J. Zool.* 72: 2016-3. DOI:10.1139/z94-273

[9] Shields, VDC. Fine structure of the galeal styloconic sensilla of larval *Lymantria dispar* (Lepidoptera: Lymantriidae). Ann. Of the Entomological Society of America 2009;102: 1116-1125. DOI:10.1603/008.102.0621

[10] Schoonhoven, LM. Plant recognition by lepidopterous larvae. Symp. Roy. Soc. Lond. 1972; 6: 87-99

[11] Shields, VDC, Mitchell, BK Responses of maxillary styloconic receptors to stimulation by sinigrin, sucrose and inositol in two crucifer-feeding, polyphagous lepidopterous species. Phil. Trans. Roy. Soc. Lond. B. 1995;347:447-457. DOI:10.1098/rstb.1995.0036

[12] Shields, VDC, Mitchell, BK The effect of phagostimulant mixtures on deterrent receptors in two crucifer-feeding lepidopterous species. Phil. Trans. Roy. Soc. Lond. B. 1995;347:459-464. DOI:10.1098/rstb.1995.0037

[13] Martin, TL Shields, VDC. Detection of alkaloids and carbohydrates by taste receptor cells of the galea of gypsy moth larvae, *Lymantria dispar* (L.) Arthropod Plant Interactions 2012;6: 519-529. DOI:10.1007/s11829-012-9209-0

[14] Martin, TL, Shields, VDC. An electrophysiolgical analysis of the effect of phagostimulant mixtures on the responses of a deterrent-sensitive cell of gypsy moth larvae, *Lymantria dispar* (L.) Arthropod Plant Interactions 2012;6: 259-267. DOI:10.1007/s11829-012-9183-6

[15] Dethier, VG, Kuch, JH. Electrophysiological studies of gustation in lepidopterous larvae. I. Comparative

sensitivity to sugars, amino acids, and glycosides. Z. Vergl. Physiol. 1971;72: 343-363. DOI:10.1007/BF00300708

[16] Schoonhoven, LM, Jermy, T. A behavioural and electrophysiological analysis of insect feeding deterrents. In:. McFarlane, N.R., editor. Crop protection agents-their biological evaluation. Academic Press, London. 1977. pp. 133-146.

[17] Dethier, VG. Mechanisms of host plant recognition. Entomol Exp. Appl. 1982;31: 49-56. DOI/10.1111/j.1570-7458.1982.tb03118.x

[18] de Boer, G. Role of bilateral chemosensory input in food discrimination by *Manduca sexta* larvae. Entomol. Exp. Appl. 1991;61:159-168. DOI:/10.1111/j.1570-7458.1991.tb02408.x

[19] Waldbauer, GP, Fraenkel, G. Feeding on normally rejected plants by maxillectomized larvae of the tobacco hornworm, *Protoparce sexta* (Lepidoptera, Sphingidae). Ann. Entomol. Soc. Am. 1961;54:477-485. DOI:10.1093/aesa/54.4.477

[20] Frazier, JL. The perception of plant allelochemicals that inhibit feeding. In: Brattsten, L.B. and Ahmad, S., editors. Molecular aspects of insect-plant associations. Plenum Press, NY. 1986. pp. 1-42.

[21] Frazier, JL. How animals perceive secondary plant compounds. In: Rosenthal, G.A. and Berenbaum, M.R., editors. Herbivores: Their Interactions with Secondary Plant Metabolites. Vol. 2. Academic Press, NY. 1992. pp 89-133.

[22] Devitt, BD, Smith, JJB. Morphology and fine structure of mouthpart sensilla in the dark-sided cutworm *Euxoa messoria* (Harris) (Lepidoptera: Noctuidae). Int. J. Insect Morphol. Embryol. 1982;11: 255-270. DOI:10.1016/0020-7322(82)90015-0

[23] Schneider, D. Insect antennae. Annual Review of Entomology. 1964;9: 103-122. DOI:10.1146/annurev.en.09.010164.000535

[24] Shields, Vonnie D.C. 1994. Ultrastructure of the aporous sensilla on the galea of larval *Mamestra configurata* (Walker) (Lepidoptera: Noctuidae). *Can. J. Zool.* 72: 2032-54. DOI:10.1139/z94-274

[25] Shields, VDC. Comparative external ultrastructure and diffusion pathways in styloconic sensilla on the maxillary galea of larval *Mamestra configurata* (Walker) (Lepidoptera: Noctuidae) and five other species. J. Morphol. 1996;228: 89-105. DOI:10.1002/(SICI)1097-4687(199604)228:1<89::AID-JMOR7>3.0.CO;2-K

[26] Broyles, JL, Hanson, FE, Shapiro, AM. Ion dependence of the tarsal sugar receptor of the blowfly *Phormia regina*. J. Insect Physiol. 1976;22: 1587-1600. DOI:10.1016/0022-1910(76)90050-0

[27] Thurm, U. Mechanoreceptors in the cuticle of the honey bee: fine structure and stimulus mechanism. Science 1964;145:1063-1065. DOI:10.1126/science.145.3636.1063

[28] Kent, KS, Hildebrand, JG. Cephalic sensory pathways in the central nervous system of *Manduca sexta* (Lepidoptera, Sphingidae). Phil. Trans. Roy. Soc. Lond. B. 1987;315:3-33. DOI: 10.1098/rstb.1987.0001

[29] Mitchell, BK, Itagaki, H, Rivet, MP. Peripheral and central structures involved in insect gustation. Micros. Res. Tech. 1999;47:401-415. DOI:10.1002/(SICI)1097-0029(19991215)47:6<401::AID-JEMT4>3.0.CO;2-7

[30] Blaney, WM, Simmonds, MSJ., Ley, SV., Jones, PS. Insect antifeedants: a behavioural and electrophysiological investigation of natural and synthetically derived clerodane

ditepenoids. Entomol Exp. Appl. 1988;46:267-274. DOI:10.1111/j.1570-7458.1988.tb01121.x

[31] Griss, C, Simpson, SJ, Rohrbacher, J, Rowell, CHF. Localization in the central nervous system of larval *Manduca sexta* (Lepidoptera: Sphingidae) of areas responsible for aspects of feeding behaviour. J. Insect Physiol. 1991;37:477-482. DOI:10.1016/0022-1910(91)90023-S

[32] Rohhrbacher, J. Fictive chewing activity in motor neurons and interneurons of the suboesophageal ganglion of *Manduca sexta* larvae. J. Comp Physiol. A. 1994;175: 629-637. DOI:10.1007/BF00199484

[33] Schoonhoven, LM, Blom, F. Chemoreception and feeding behaviour in a caterpillar: towards a model of brain functioning in insects. Entomol. Exp. Appl. 1988;49:123-129. DOI:10.1111/j.1570-7458.1988.tb02483.x

[34] Fraenkel, GS. The raison d'être of secondary plant substances. Science 1959;129:1466-1470. DOI: 10.1126/science.129.3361.1466

[35] Bernays, EA, Chapman, RF. Host-plant selection by phytophagous insects. Chapman Hall, New York. 1994. DOI:10.1007/b102508

[36] Schoonhoven, LM. Chemical mediators between plants and phytophagous insects. In: D.A. Nordlund, R.L. Jones, and W.J. Lewis editors. Semiochemicals: their role in pest control. John Wiley, New York; 1981. pp. 31-50.

[37] Whittaker, RH. The biochemical ecology of higher plants. In: E. Sondheimer, and J.B. Simeone editors. Chemical ecology. Academic Press, New York, NY; 1970. pp. 43-70. DOI:10.1016/B978-0-12-654750-4.50009-8

[38] Schoonhoven, LM. Secondary plant substances and insects. Rec. Adv. Phytochem. 1972;5:197-224. DOI:10.1016/B978-0-12-612405-7.50013-8

[39] Wink, M. Plant secondary metabolites modulate insect behavior-steps toward addiction? Front. Physiol. 2018;9:364. DOI:10.3389/fphys.2018.00364

[40] Wink, M., Schimmer, O. Molecular modes of action of defensive secondary metabolites. In: M. Wink editor. Functions and biotechnology of plant secondary metabolites. Blackwell, Oxford; Annual Plant Reviews. 2010;39:21-161. DOI:10.1002/9781444318876

[41] Jermy, T. Feeding inhibitors and food preference in chewing phytophagous insects. Entomol. Exp. Appl. 1966;9:1-12. DOI:10.1111/j.1570-7458.1966.tb00973.x

[42] Wink, M. Physiology of accumulation of secondary metabolites with special reference to alkaloids, In: F. Constabel, and I.K. Vasil, editors. Cell cultures and somatic cell genetics of plants, vol. 4. Academic Press, San Diego, CA. 1987; pp. 17-42.

[43] Shields, VDC, Broomell, BP, Salako, JOB. Host selection and acceptability of selected tree species by gypsy moth larvae, *Lymantria dispar* (L.). Ann. Entomol. Soc. Am. 2003;96: 920-926. DOI:10.1603/0013-8746(2003)096[0920:HSAAOS]2.0.CO;2

[44] Mosher, F H. Food plants of the gypsy moth in America. U.S.D.A. Bull. No. 250; 1915. pp. 1-39.

[45] Shields, VDC, Mitchell, BK. Sinigrin as a feeding deterrent in two crucifer-feeding, polyphagous lepidopterous species and the effects of feeding stimulant mixtures on deterrency. Phil.

Trans. Roy. Soc. Lond. B. 1995;347:439-446. DOI:10.1098/rstb.1995.0035

[46] Shields, VDC, Rodgers, EJ, Arnold, NS, Williams, D. Feeding responses to selected alkaloids by gypsy moth larvae, *Lymantria dispar* (L.). Naturwissenschaften 2006;93:127-130. DOI:10.1007/s00114-005-0070-1

[47] Shields, VDC, Smith, KP, Arnold, NS, Gordon, IM, Shaw, TE, Waranch, D. The effect of varying alkaloid concentrations on the feeding behavior of gypsy moth larvae, *Lymantria dispar* (L.) (Lepidoptera: Lymantriidae). Arthropod-Plant Interactions 2008;2:101-107. DOI:10.1007/s11829-008-9035-6

[48] Shields, VDC, Martin, TL. The Effect of Alkaloids on the Feeding of Lepidopteran Larvae. In: Cassiano, Nicole M., editor. Alkaloids: Properties, Applications and Pharmacological Effects, Nova Science Publishers, Inc. Hauppauge, NY. Chapter 6, 2010. pp. 109-138.

[49] Wink, M. Allelochemical properties or the raison d'être of alkaloids, In: Cordell, GA, editor. The alkaloids: chemistry and pharmacology, vol. 43. Academic Press, Inc. Boston, MA. 1993. pp. 1-118. DOI:10.1016/S0099-9598(08)60134-0

[50] Wink, M, Theile, V. Alkaloid tolerance in *Manduca sexta* and phylogenetically related sphingids (Lepidoptera: Sphingidae). Chemoecology 2002;12: 29-46 DOI:10.1007/s00049-002-8324-2

[51] Doskotch, RW, El-Feraly, FS, Fairchild, EH, Huang, C. Isolation and characterization of peroxyferolide, a hydroperoxy sesquiterpene lactone from *Liriodendron tulipifera*. J. Org. Chem. 1977;42: 3614-3618. DOI:10.1021/jo00442a037

[52] Miller, JS, Feeny, P. Effects of benzylisoquinoline alkaloids on the larvae of polyphagous Lepidoptera. Oecologia (Berl.). 1983;58: 332-339. DOI:10.1007/BF00385232

[53] Barbosa P, Krischik VA. Influence of alkaloids on feeding preference of eastern deciduous forest trees by the gypsy moth *Lymantria dispar*. Am Nat. 1987; 130: 53-69. DOI:10.1086/284697

[54] McCormick, A., Arrigo, L, Eggenberger, H, Mescher, MC, De Moraes, CM. Divergent behavioural responses of gypsy moth (*Lymantria dispar*) caterpillars from three different subspecies to potential host trees. Sci. Rep. 2019; 9: 8953. DOI:10.1038/s41598-019-45201-3.

[55] Schoonhoven, LM, Blom, F. Chemoreception and feeding behaviour in a caterpillar: towards a model of brain functioning in insects. Entomol. Exp. Appl. 1988;49:123-129. DOI:10.1111/j.1570-7458.1988.tb02483.x

[56] Glendinning, JI, Chaudhari, N, Kinnamon, SC. Taste transduction and molecular biology. In Finger, T.E., Silver, W.L., Restrepo, D. editors. The neurobiology of taste and smell. Wiley-Liss, Inc., NY. 2000. pp. 315-351.

[57] Hodgson, ES, Lettvin, JY, Roeder, KD. Physiology of a primary chemoreceptor unit. Science 1955;122:417-418. DOI:10.1126/science.122.3166.417-a

[58] Schoonhoven LM, van Loon, JJA. An inventory of taste in caterpillars: each species its own key. Acta Zool. Hung. 2002;48: 215-263.

[59] van Loon J.J.A. Chemosensory basis of feeding and oviposition behaviour in herbivorous insects: a glance at the periphery. In: Städler E., Rowell-Rahier M., Bauer R. editors. Proceedings of the 9th International Symposium on Insect-Plant Relationships. Series Entomologica, vol 53. Springer, Dordrecht. 1996. Pp. 7-13. DOI:10.1007/978-94-009-1720-0_2

[60] Städler, E. Contact chemoreception. In: Bell,WJ, Cardé, RT, editors. Chemical Ecology of Insects. Chapman and Hall, New York. 1984, pp. 3-35. 10.1007/978-1-4899-3368-3

[61] Dethier, VG, Crnjar, RM Candidate codes in the gustatory system of caterpillars. J. Gen. Physiol. 1982;79:549-569. DOI:10.1085/jgp.79.4.549

[62] Schoonhoven, LM Biological aspects of antifeedants. Entomol. Exp. Appl. 1982:31:57-69 DOI:/10.1111/j.1570-7458.1982.tb03119.x

[63] Schoonhoven, LM, Blaney, WM, Simmonds, MSJ. Sensory coding of feeding deterrents in phytophagous insects. In: Bernays, EA, editor. Insect-plant interactions. Vol. 4. CRC Press, Boca Raton, FL. 1992, pp. 59-79.

[64] Schoonhoven, LM. What makes a caterpillar eat? The sensory coding underlying feeding behavior In Chapman, RF, Bernays, EA, Stoffolano, JG, editors, Chemoreception and Behavior. Springer-Verlag, New York. 1987), pp. 69-97,

[65] Schoonhoven, LM, Jermy, T, van Loon, JJA. *Insect-plant biology. From Physiology to evolution.* Chapman and Hall, London; 1998

[66] Blaney, WM, Simmonds, MSJ, Ley, SV, Jones, PS Insect antifeedants: a behavioural and electrophysiological investigation of natural and synthetically derived clerodane ditepenoids. Entomol Exp. Appl. 1988;46:267-274. DOI/10.1111/j.1570-7458.1988.tb01121.x

[67] Hanson, FE, Peterson, SC Sensory coding in *Manduca sexta* for deterrence by a non-host plant, *Canna generalis*. Symp. Biol. Hung. 1990;39: 29-37. DOI:10.1111/j.1365-3032.1993.tb00601.x

[68] Yarmolinsky, DA, Zuker, CS, Ryba, NJP. Common sense about taste: from mammals to insects. Cell 2009;139: 234-244. DOI:10.1016/j.cell.2009.10.001

[69] Kasubuchi, M, Shii, F, Tsuneto, K, Yamagishi, T, Adegawa, S, Endo, H, Sato, R. Insect taste receptors relevant to host identification by recognition of secondary metabolite patterns of non-host plants. Biochem. Biophys. Res. Commun. 2018;499: 901-906. DOI:10.1016/j.bbrc.2018.04.014

[70] Zhang, Z,-J, Zhang, S-S, Niu, B-L, Ji, D-F, Liu, X-J, Li, Mu-Wang, Bai, H, Palli, SR, Wang, C-Z, Tan, A-J. A determining factor for insect feeding preference in the silkworm, *Bombyx mori*. PLoS Biol.